A CMOS Self-Powered Front-End Architecture for Subcutaneous Event-Detector Devices

Jordi Colomer-Farrarons
University of Barcelona
Electronics Department
Marti i Franques 1
08028 Barcelona
Spain
jcolomer@el.ub.es

Pere Lluís Miribel-Català
University of Barcelona
Electronics Department
Marti i Franques 1
08028 Barcelona
Spain
pmiribel@el.ub.es

ISBN 978-94-007-0685-9 e-ISBN 978-94-007-0686-6
DOI 10.1007/978-94-007-0686-6
Springer Dordrecht Heidelberg London New York

Library of Congress Control Number: 2011922058

Printed on acid-free paper

Springer is part of Springer Science+Business Media (www.springer.com)

Jordi Colomer-Farrarons · Pere Lluís Miribel-Català

A CMOS Self-Powered Front-End Architecture for Subcutaneous Event-Detector Devices

Three-Electrodes Amperometric Biosensor Approach

Springer

Preface

This book presents the conception and prototype realization of a Self-Powered architecture for subcutaneous detector devices. The architecture is designed to work as an event detector or threshold level alarm of some substances, ions, etc ... that are detected through three-electrodes amperometric Biosensor approach.

In this book, the final device is envisaged as a Low-Power subcutaneous implantable application powered by an inductive link, one emitter antenna at the external side of the skin and the receiver antenna under the skin.

The sensor is controlled with a potentiostat circuit and then, a post-processing unit detects the desired levels and activates the transmission via a backscattering method using the inductive link. All the instrumentation, except the power module, is implemented in the BioChip ASIC.

Following the idea of the powering link to harvest energy of the magnetic induced link at the implanted device, a Multi-Harvesting Power Chip (MHPC) has been also designed. This power management unit, able to collect energy from three different sources, solar, mechanical vibrations and magnetically induced energy, converts the energy of the environment into usable electrical energy. In that way, any type of Low-Voltage Low-Power electronics can be powered. This MHPC is used as a supply unit in the designed subcutaneous device.

This book is based on the research carried out in the Smart Self-Powered BioInspired Systems (S^2PBS) of the SIC – BIO Group at the Universitat de Barcelona, Spain, focused on developing discrete-to-integrate electronics for specific instrumentation.

This book is divided in five chapters:

- In Chapter 1, an overview of the state of the art of energy harvesting is done. Several ambient power sources and human body-centered sources are commented. Then, a general introduction to Biosensors and electronics for three electrodes sensors is provided.
- Chapter 2 discuses the design of an ASIC able to recollect energy from several different energy harvesting power sources. All circuits are theoretically and experimentally described and validated. The final IC, called MHPC, is tested and validated.

- Chapter 3 focuses on the design of the integrated instrumentation IC (BioChip) to work with the three electrodes amperometric sensor. The potentiostat amplifier is described in detail and validated through experimental results using a real electrochemical substance. Furthermore, the conception of a Lock-In amplifier is also introduced in this chapter as well as the transcutaneous link data transmission.
- Finally, the Self-Powered front-end architecture for subcutaneous detector devices is presented in Chapter 4. The device combines the MHPC, BioChip, communication antenna, etc... in the same substrate that is fully validated through different test.
- Chapter 5 summarizes the main contributions presented in this book. Moreover, some possible future aspects that may arise from this work are presented.

Acknowledgments

We would like to acknowledge the significant contributions and successful discussions of Prof. Dr. Marise Bafleur, CNRS Research Director Responsible of the "Intégration des Systèmes de Gestion de l'Energie" (ISGE) group at the LAAS Laboratory in Tolouse, France. An special mention to Prof. Dr. Teresa Riesgo "Tere", head of Center of Industrial Electronics (CEIP-UPM) at the Polytechnic University of Madrid, for her valuable comments and support through the elaboration of all this work.

Our most sincere thanks to Prof. Dr. Josep Samitier Martí, associate director of the IBEC institute (Institute for Bioengineering of Catalonia), head of the IBEC - NanoBioEngineering Group and SIC – BIO Group at the University of Barcelona, and coordinator of the Spanish Platform on Nanomedicine, to give us the opportunity to work in the SIC – BIO group and for his support on making this research possible.

A special mention to Dr. Martin Arundell of the School of Biological Science, University of Southampton, for his "hardware" support and his essential comments and contributions in the understanding of the electrochemical essentials. Also, all our gratitude to the people of IMEC-INVOMEC, specially to Jorgo Tsouhlarakis and Steven Redant.

We would like to express our deepest gratitude to our close families and friends. They have always helped us in all the important moments and have supported all our decisions.

Contents

Abbreviations

AC	Alternating current
AM	Amplitude modulation
CGC	Clock generation circuit
CMOS	Complementary metal-oxide semiconductor
CV	Cyclic voltammetry
DC	Direct current
EIS	Electrochemical impedance spectroscopy
ESR	Equivalent series resistor
LDO	Low dropout
LPF	Low-pass filter
MOSFET	Metal-oxide-semiconductor field-effect transistor
MSD	Multiple storage device
OTA	Operational transconductance amplifier
PCB	Printed circuit board
POR	Power-On-Reset
RF	Radio frequency
RFID	Radio frequency identifier
SDG	Synchronous demodulated channel
SMD	Surface mount device
SSD	Single storage device
TG	Transmission gate
TIA	Transimpedance amplifier

Chapter 1
Introduction

Abstract This chapter introduces the state-of-the-art in the main topics covered in the book: energy harvesting, with special interest in the body energy harvesting sources, biosensors, and finally the electronics for them. These are the main aspects to consider in the envisaged conception of the Self-Powered architecture for subcutaneous detector device.

Keywords Low-power instrumentation · Miniaturized biomedical system · Wireless implanted devices · Self-Powered devices · Event detector implantable devices

The integration of medicine and electronic technologies allows the development of biomedical devices able to diagnose and/or treat pathologies by detecting and/or monitoring pathogens, multiple ions, PH changes, and so on. Moreover, the advances in different areas such microelectronics, microfluidics, microsensors and biocompatible materials open the door to develop human body Lab-on-Chip Implantable Devices and Point-of-Care in-vitro devices.

The key aspect on the development of subcutaneous applications consists on combining the powering and the instrumentation in the same device, hence two main problems should be overcome. The first one consists on the integration of the necessary instrumentation and communication electronics to control the sensors/Biosensors and to send the information through human skin, whereas the second one resides in the way to transfer enough energy to power the device.

Some examples of implantable telemetry electronics integration can be found in the literature like in [1, 2] or [3] and in [4] and [5] where two integrated electronics for in-vivo monitoring are introduced. Although many efforts have been focused on developing robust electronics for in-vivo monitoring, that is continuously sensing, recording and data transmission, not all the in-vivo subcutaneous applications require these circuits. There are other less studied applications more focused on the detection of presence or absence of a certain levels of proteins, antibodies, ions, oxygen, glucose, etc…

These in-vivo subcutaneous event detection circuits or True/False applications [4] work as an alarm; when the analyzed concentration level exceed, under or over,

J. Colomer-Farrarons, P.L. Miribel-Català, *A CMOS Self-Powered Front-End Architecture for Subcutaneous Event-Detector Devices*, DOI 10.1007/978-94-007-0686-6_1,

a threshold value or the system detects the presence of zero tolerance pathogens like salmonella [4], the systems send to an external reader a signal indicating the fault or showing the value as in some pregnancy tests.

For instance, in the case of glucose monitoring, the detection of a threshold decrease in the glucose level it is mandatory to avoid critic situations like the hypoglycaemia [6]. In that case, the True/False implantable device could check the glucose levels to manage an automatic method to inject insulin to regulate the glucose level on blood.

The second question to be considered is how supply energy to the subcutaneous or implantable devices. Since it is quite difficult to use the classical bulky batteries or other more modern type of batteries under the human skin, the device is not a pacemaker that is inserted into the chest, this is implanted just under the skin, a smart solution consists in harvesting the energy from the environment and, in that particular case, to recollect the energy coming from the human being to self-power the implanted device.

In application areas as diverse as aircraft construction, personal health care systems or burglary detection systems, energy harvesting is on the verge of developing a market with a multibillion dollar potential. For instance, an electronic device including a semiconductor-based thermo electric generator (TEG) that transforms temperature differences into electrical energy could be worn on the body using the human body heat as an energy source for medical monitoring. Furthermore, the total amount of energy demanded to the energy suppliers could be strongly reduced just walking around while carrying on the body or on the clothes several autonomous portable devices in which energy harvesting from vibrations has been included.

The benefits of this green local actuation could get a huge impact in the global energy demand. In the particular case of personal healthcare, the combination of harvesting sources and/or thin-film batteries opens new fields of interest, like Patient Monitoring & Tracking, Transdermal Drug Delivery, Accelerated Healing Devices and Pharmaceutical Compliance in the conception of a Body Sensor Network (BSN).

The miniaturization of electronics and the combination with solutions like the thin-film batteries by Silicore [7] are interesting options in the development of new Transdermal Drug Delivery (TDD) methods [8, 9] with the idea to eliminate needles, Fig. 1.1, and to ensure proper dosage control, closing the feedback of an artificial pancreas.

Moreover, an inductive coupling RF power harvesting is a growing alternative method to transmit energy transcutaneously instead of batteries or wires [10–12], and for the particular case of subcutaneous devices it seems a good approach.

Furthermore, this alternative permits to establish a bidirectional communication between the implanted device and an external base or reader. Some implantable RF circuits based on inductively coupling can be found in the literature like in [13, 14]. Several energy sources as well as the state of the art in energy harvesting are introduced later on in this chapter, whereas on Chapter 2 is developed our approach to Multi-Harvesting Power Chip (MHPC) applicable to generic developments.

A the end of this work, on Chapter 4, it is presented a generic implantable front-end architecture based on an inductive coupling for in-vivo presence or absence

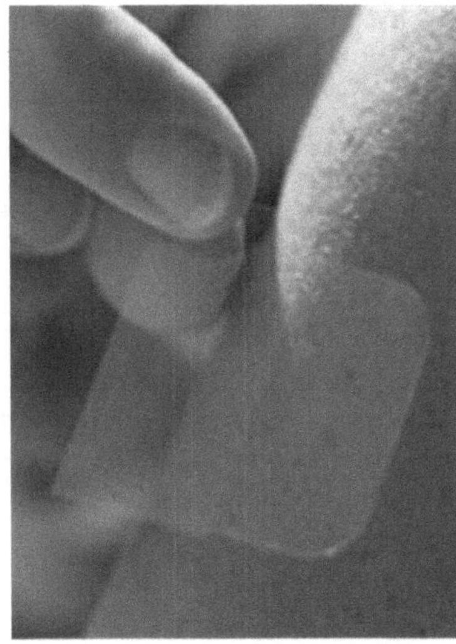

Fig. 1.1 Patch example from [8]

detection of pathogens, ions, oxygen concentration, etc... for a generic amperometric Biosensors application. The proposed architecture includes the on-chip biasing, the potentiostat to drive the sensor, and the modulation and communication block.

The final step of this book is the conception of a first capsule prototype with an ideal size less than 4.5 cm long and 2.5 cm in diameter following the same philosophy like some contraceptives subcutaneous implantable devices like Norplant®, Jadelle® or Implanon® [15, 16].

All circuits in this work have been developed with the 130 nm technology from ST Microelectronics available through the CMP Tima. This tech. gives a nice tradeoff between area and Analog/Digital performances for that type of applications.

1.1 Energy Harvesting in Human and Non-human Activities

1.1.1 Introduction

Nowadays, there is an enormous interest on renewable energy sources and its applications, especially for high-power levels [17, 18, 19]. There is also an increasing interest in using free available external energy sources for powering small electronic systems, a process known as energy harvesting [20]. However, the amount of energy that can be obtained from these external ambient sources like vibrations, heat gradients, light, radio waves [21–23, , , 14], or from human activity [25, 26] is limited regarding power consumption levels.

Some already published works show the design of self-powered systems based on MEMS micro-power sources like electromagnetic micro-power MEMS generators [27, 28], variable capacitors [29], piezoelectric-based generators [30, 31], or defining completely an integrated sensor node [32]. Examples of these technologies are the design of self-powered wireless networks [33] formed by distributed sensors which are capable of monitoring the real-time condition of motors, turbines, pumps, and gear boxes, or the design of heating or ventilation control systems for burglary alarms.

One of the most interesting application fields is biology where self-powered RFID bio tags can monitor the temperature, pressure [34, 35] or check animal's healthcare [36, 37].

The continuous advances in the semiconductor's integration technology related to the reduction of the transistor's size allow the industry to develop these new self-powered portable electronic devices which usually include in their System-in-Package (SiP) or System-on-Chip (SoC) a great variety of circuitry and functions like wireless sensor networks or biomedical electrical instrumentation. However, the main problem with these new devices resides in the power supply system, which is usually based on a typical bulky battery solution often with much larger size than the SiP or SoC itself.

The size, width, finite energy lifetime, and battery replacement are important drawbacks for those small and portable applications that need long lifetime energy supply systems [38–40]. So, novel and innovative energy supply alternatives must be explored to remove or replace the battery dependence and make feasible the deployment of these batteryless devices. Focusing on this field, a new trend in the research of energy sources for low-power applications is growing rapidly over the last years.

This approach consists in harvesting the available energy of the environment in order to supply enough power to the electronic applications instead of using a battery or other technologies with short lifetime and finite amount of energy.

1.1.2 State-of-the-Art of the Energy Harvesting

The first energy cell, the battery, was invented by Alessandro Volta 1799. His invention predated the first mechanical to electrical energy converter device discovered by Michael Faraday. At the beginning, batteries were the only way to generate electrical energy. It was at the end of nineteenth century when electricity arrives to the cities by wires and the batteries were relegated to mobile platforms.

Battery technology underwent a tremendous progress since it was first discovered. This progress enabled the explosion of a huge number of new applications such for mobile devices. However, new challenges in technology added to some intrinsic limitations of batteries have motivated the research to look for new smart energy generation solution systems.

These new smart energy systems are conceived to actively "monitor and optimize themselves". The requirements and expectations for these systems are: efficiency, sustainability, intelligence, autonomy, a high degree of integration, etc... These expectations rule out the use of batteries as energy sources because they have a limited life time, which is in contrast to the concept of "Smart". In addition, batteries produce pollutant residues. New energy solutions must have a very long life time and must be sustainable.

A very interesting publication to the "Economist Group" about future trends was presented in 2003 by the Microsoft chairman and chief software architect Bill Gates [41]. The publication was named "The Disappearing Computer". In that publication, he envisioned a future world where ubiquitous embedded systems will interact with people whether they know it or not. Those envisioned systems will have to run without batteries and use power harvesting systems to transform environmental energy into usable electrical energy.

> As people find more ways to incorporate these inexpensive, flexible and infinitely customizable devices into their lives, the computers themselves will gradually "disappear" into the fabric of our lives. We are still a long way from a world full of disembodied intelligent machines, but the computing experience of the coming decade will be so seamless and intuitive that–increasingly–we will barely notice it. At the same time, computing will become widespread enough that we will take it for granted–just as most people in the developed world today trust the telephone service.
>
> The pervasiveness and near–invisibility of computing will be helped along by new technologies such as cheap, flexible displays, fingernail-sized MEMs (microelectromechanical systems) chips capable of storing terabytes of data, or inductively powered computers that rely on heat and motion from their environment to run without batteries.

Thus, the solution for this energy problem seems to pass through the design of systems capable of continuously transform energy from the environment into electrical energy useful to supply an electronics device. This procedure is known as "power harvesting". Basically, four different power sources are available in the environment: mechanical vibrations, solar, thermal and Magnetically Induced.

1.1.2.1 Mechanical Vibrations

Mechanical vibrations can be found in many places. In floors and walls of a building caused by nearby machinery. In automobile chassis or tires, in jet engine housing, human walking, etc...

These vibrations present a wide range of amplitudes and frequencies. Medium frequency vibration appears in our daily life. For example, inside a building or inside a lab, there are different vibrations produced by different sources. Figure 1.2 shows typical vibration sources that act inside a laboratory [42].

Moreover, in our daily home life it is also used some machinery that also produces vibrations like a microwave oven or a milling machine. In general, we are submitted to vibrations from 1 to 100 Hz. Some other intermediate vibrations appear with higher accelerations in vehicles, i.e. it is possible to find vibration frequencies up to 1 kHz @ 10m/s^2 inside a car. Higher frequencies can be found in industrial applications. Some specific engines such as some turbines or motors generate frequencies > 1 kHz.

Researchers usually have long designed systems exploiting the oscillations of a proof mass to harvest mechanical vibration energy. These kinds of converters are also known as "Inertial Converters". There are different ways to convert this mechanical energy into electrical energy. In fact there exist three kinds of physical

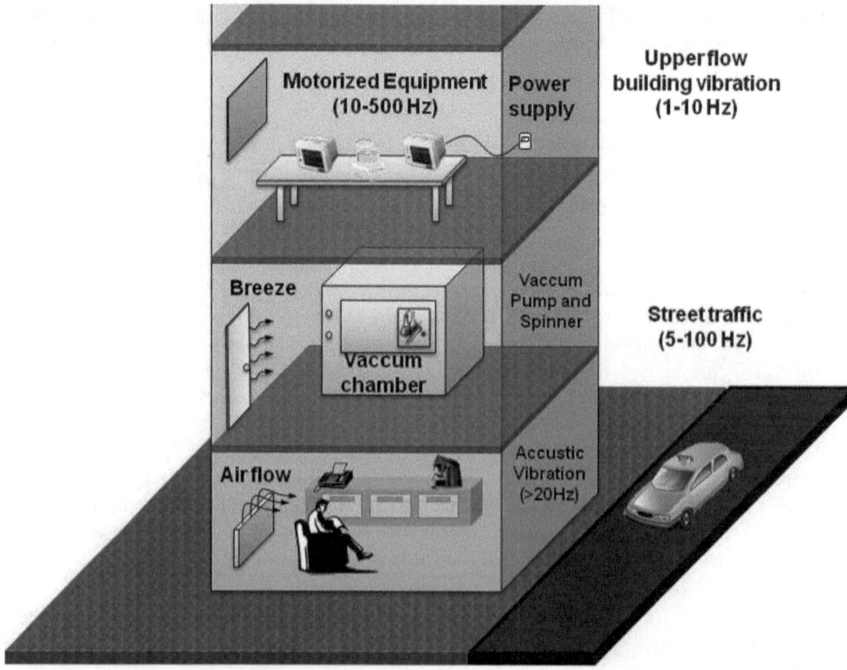

Fig. 1.2 Common environmental sources of noise in a laboratory [43]

principles that can be used: (1) magnetic induction converters, (2) electrostatic converters and (3) piezoelectric converters.

The magnetic induction transducer (1) is based on Faraday's law. It is based on the motion of an electric conductor through a magnetic field. Typically the conductor is wound in a coil to make an inductor. The relative motion between the coil and magnetic field causes a current flowing in the coil.

Some relevant research groups have been designing electromagnetic power generators. MIT researchers [44] developed a generator able to deliver 400 μW using the human walking activity as vibration source. On the other hand, other group from Sheffield, presented a 1 mm^3 size electromagnetic micro-generator, able to generate 0.3 μW from a 4 MHz input excitation.

Electrostatic generation (2) consists of two conductors separated by a dielectric (i.e. a capacitor), which moves relative to one another. As the conductors move, the energy stored in the capacitor changes, thus providing the mechanism for mechanical to electrical energy conversion. Menninger et al. of MIT presented an electrostatic generator that employs a variable micro machined capacitor to recollect ambient vibrations.

The third physical technique consists in using the well known piezoelectric materials. These materials generate an electric field or electric potential in response to a mechanical stress. Like the other techniques it is possible to find some examples of generator based on piezoelectric like in [42].

There are a lot of variables and trade-offs to take into account when designing systems for harvesting from vibrations. Some of the more important ones are: volume restrictions, specifications of the vibration source and environment of application, the available electronic and mechatronic technology, etc... Although all these aspects will influence the solution to be adopted, it is interesting to make a quantitative comparison in terms of energy density inherent to each type of converter.

Table 1.1 presented by Roundy et al. [45] shows the expressions for the three energy converters. Moreover, a computation of the theoretical and practical maximum values is shown. The piezoelectric energy density equation is obtained considering the energy density in a dielectric material and the electric field induced by a mechanical stress. The energy density equation for an electrostatic converter is directly computed as the energy density in a dielectric material and finally, the energy density in the electromagnetic generator is directly computed as the energy of a magnetic field.

Table 1.1 Summary of maximum energy density of three types of mechanical transducers

Type	Governing equation	Practical max. (mJ/cm^3)	Theoretical max.
Piezoelectric	$u = \sigma_y^2 k^2 / 2Y$	17.7	335 mJ/cm^3
Electrostatic	$u = 0.5\varepsilon E^2$	4	44 mJ/cm^3
Electromagnetic	$u = B^2 / 2\mu_0$	4	400 mJ/cm^3

The quantitative comparison is based on the evaluation of the maximum energy densities that these materials can sustain. In the case of the piezoelectric converter, the maximum energy density is achieved when the material sustains the maximum stress before breaking; this is named the "yield strength".

Table 1.1 also shows that piezoelectric converters have higher practical energy densities than the other two options. Nevertheless, it is important to have in mind other qualitative and performance comparisons for these converters.

- *Electromagnetic* converters produce very low voltages. This is a serious problem because the signal generated from the converter is an AC signal so it needs to be rectified. In order to increase this signal a heavy transformer has to be used so the dimensions of the whole system increase. These converters have other problems intrinsically associated to them. The role of the magnetic fields in the transduction process can produce interferences with its associated electronic circuitry. These converters have the advantage of being easily micro-mechanized.
- *Electrostatic* converters have the disadvantage of requiring an external power source to charge the capacitor beforehand; moreover they need a mechanical stopper in order to avoid the contact between the two plates. This mechanical stopper will increase the damping of the system. On the other hand, these converters produce higher voltages than the electromagnetic ones, and can be directly rectified. A very interesting advantage of these converters is their easy integration

in micromechanics systems that allow the design of compact and specifically designed systems.

- *Piezoelectric* converters have the highest energy density. The voltage and current levels are in an acceptable range and are easily modified to fulfill the input specifications. These converters do not need any magnetic field or mechanical stopper.

1.1.2.2 Solar Energy

It consists on using the energy radiated by the sun and it is widely used as a power source in many applications. In fact, solar radiation varies around the world due to weather conditions, longitude and latitude, etc. However, it is possible to give an averaged value for a direct bright sun of about 100 mW/cm^2. It is also possible to use the emitted light inside a building; in this case an average value of about 100 μW/cm2 is expected [46].

The conversion principle for changing light into electrical energy is based on the photovoltaic effect. This effect is the result from the interaction between light and matter, the absorption of the photon energy by the photovoltaic cell generates electrical current. Silicon solar cells are a mature technology for a large scale.

Photovoltaic systems are found from the Megawatt to the milliwatt range, producing electricity for a wide range of applications with efficiencies from 12 to 25%. There are thousands of companies that supply silicon solar cells and spend millions of dollars to improve its efficiencies while reducing costs.

However, new materials with tremendous challenging potential applications are being used to form photovoltaic cells. Nanosolar [47], Konarka [48], and other companies are developing energy-producing material with solar cells embedded in thin sheets of plastic, which promises to drastically reduce solar power's cost. This "power plastic" can be laminated onto any surface, from rooftops to laptops to automobiles [49].

1.1.2.3 Thermoelectric Generators

The use of heat gradients to generate energy was one of the first ways to accomplish energy conversion. The classical "vapor machines" used the expansion of a gas, that when heated, moved a piston by pressure and generates a mechanical motion. However, since it would need to combust some kind of substance or material, this kind of systems could not be considered as an energy harvesting system,

On the other hand, in the environment, there are different constant sources of wasted heat that can be used to generate electrical power. The human body is always at the same temperature even the environmental temperature changes, this heat gradient can be used to generate electrical current.

There are different methods or principles to convert a heat gradient into electrical energy: 1) thermoelectric generators and 2) pyroelectric generators.

Thermoelectric generators (1), which have been of great interest in the recent years, are based on a junction of two dissimilar materials "n" and "p" semiconductors. These are connected by a metal conductor at the hot and at the cold side of the gradient. Temperature gradients across the thermoelectric material drives electron charge carriers from the hot to the cold junction and produce a voltage. Thermoelectric generators are a good choice since they can be microfabricated with classical microfabrication techniques.

In 1997, Stordeur et al. designed a micropower thermoelectric generator able to produce 20 μW with $\Delta T = 20$ °K [50]. Nowadays, some companies are selling products based on thermoelectric generators. For example, Seiko placed for sale a thermal wristwatch that uses 10 thermoelectric modules to generate sufficient microwatts to run its mechanical clock [51]. Applied Digital Solution presented the Thermo Life that is a thermoelectric generator measuring 0.5 cm^2 and 1.6 mm thick, which can supply energy to a Biosensor when it contacts with the skin. It is able to supply 5 μA @ 3 V with a 5°C of gradient temperature, [52].

Pyroelectric solutions (2) are based on polar materials when they are subjected to temperature changes. The effect is similar to the piezoelectric effect but, instead of electric charges appearing when the material is deformed, superficial charges appear in a certain direction, when the material changes its temperature.

1.1.2.4 Mechanical, Thermoelectric and Solar Energy Comparison

A comparison of the different energy harvesting technologies is extracted from the analysis made by Shad Roundy in his dissertation submitted at the Virginia Polytechnique Inststute [53]. Table 1.2 shows the energy densities for a set of different environmental scavenged power sources and for some energy reservoirs.

The table is composed by three columns. The first one shows the energy density during the first year. The second column shows the energy density for the first year and up to the tenth year of life. This information is given to show the decrease in the performances of the energy reservoirs after several uses. The third column shows the source of information.

While this table of comparison is by no means exhaustive, it does provide a broad overview of potential methods to scavenge energy and energy storage systems. It is clearly seen that the energy reservoir undergoes to a considerable reduction of its energy density (one order of magnitude) after the first years of use, except for the nuclear uranium based sources.

This degradation in performance reduces the number of applications where this energy sources could be used. In contrast, the harvesting sources theoretically present the same performance after 1 year of life; therefore, it is possible to think in applications where the systems are completely autonomous for a long time.

Comparing the different sources, it is clear that the most energetic is the solar solution in a sunny day. This is the most widely used solution due to the ubiquitous of solar light. Mechanical vibrations are presented as the next more energetic solution because mechanical vibrations surround our daily life. Furthermore, some

Table 1.2 Comparison of energy scavenging and energy storage methods. Note that leakage effects are taken into consideration for batteries [53]

		Power density (μW / cm^3) 1 Year lifetime	Power density (μW / cm^3) 10 Year lifetime	Source
Scavenged power source	Solar (outdoors)	15,000 – direct sun 150 – cloudy day	15,000 – direct sun 150 – cloudy day	Commonly available
	Solar (indoors)	6 – office desk	6 – office desk	0
	Vibrations	200	200	Roundy 2002
	Acoustic noise	0.003 @ 75 dB 0.96 @ 100 dB	0.003 @ 75 dB 0.96 @ 100 dB	Theory
	Daily temp. variation	10	10	Theory
	Temp. gradient	15 @ 10 °C gradient	15 @ 10 °C gradient	Stordeur and Stark 1997
	Shoe inserts	330	330	Shenk and Paradiso 2001
Energy reservoirs	Batteries (non-recharge lithium)	45	3.5	Commonly available
	Batteries (rechargeable Lithium)	7	0	Commonly available
	Hydrocarbon fuel (micro heat engine)	333	33	Mehra 2000
	Fuel cells (methanol)	280	28	Commonly available
	Nuclear isotopes (uranium)	6×10^6	6×10^5	Commonly available

theoretical and experimental implementations demonstrate generations of about 200 μW/cm^2.

Big efforts are being done to accomplish designs for harvesting this energy. The last raw of the scavenged solutions shows the shoe insert solution. Some researchers have pointed out the possibility to use the energy damped in the sole of a shoe to drive some kind of wearable device. It is possible to harvest around 300 μW/cm^2 of power density, for normal walking conditions. The table shows other sources however the amount of energy these sources provide is very small.

1.1.2.5 Magnetically Induced Energy

Nowadays, wireless communications and transmissions are present in our daily activities generating lots of energy in electromagnetic form. Although, it is not strictly a power generator that can be easily and naturally found in the environment, the electromagnetic radiation can be used to power small devices such as energy transport media. Following that idea, terms like Witricity [54] or Wireless Power Transmission or Inductive Powering are becoming popular to power some equipment.

Wireless power transmission is basically based on the use of two electromagnetic resonant objects (i.e. one transmitter and one receiver antennas) tuned at the same frequency. In that way, the energy is transferred thanks to the electromagnetic coupling between both objects. Moreover, an electromagnetic field can be specifically designed to work at a specific frequency and orientation.

The possibility of industrial applications of this Wireless Power was demonstrated by MIT Researchers [55, 56]. They power a 60 W light bulb using two cooper coils of 60 cm diameter working at 9.9 MHz. The coils were inductively coupled achieving an efficiency of 45% for a separation of 2 m, one working as emitter and the other connected to the bulb.

Moreover, in the biomedical field, the use of an Inductive Link working at the appropriate frequency is a valuable option to transmit energy transcutaneously. Thus, the subcutaneous implantable devices can be designed avoiding the use of bulky batteries or even wires through the human skin. Other advantage is the possibility to establish a bidirectional data communication over the same Inductive link used for energy transmission.

This smart combination of power and transmission makes the use of Magnetic Induction highly valuable for sensitive applications like human subcutaneous implantable devices.

A widely used example of inductive links is a microchip implant for pet's identification. It is a Radio Frequency Identifier tag (RFID) with a unique chip number. This chip is implanted into the pet, cat, dog, etc... and the number is automatically associated to the owner after filling an enrolment form with the chip number, owner name and contact information. In that way, if the pet is lost or stolen, it can be scanned to obtain the owner information.

1.1.3 Body Harvesting

At that point, after study different environmental energy sources, it is time to analyze the energy harvesting sources available from the human body. Several studies and analysis have been reported that the human body can be used as a generator. Basically, the body could generate power in two ways, passively or actively.

Passive human generation of power for a product means that the person does not perform any special activity to drive the product. That is, passive powering takes place when the user does not have to do any task different to normal tasks associated with the product.

On the other side, active generation means that the person performs an activity that when using the same product not powered by a human would not have been performed. Generally speaking, more power can be generated actively than passively [21].

There are different available sources to harvest energy in the body, thermal, vibration or even magnetically induced thanks to the advances in portable local wireless systems. The human body is a tremendous storehouse of energy but, some key aspects must be taken into account.

The first source approach is related to the body heat. Since the human body emits energy as heat, it follows naturally to try to harness this energy. However, Carnot efficiency puts an upper limit on how well this waste heat can be recovered. Assuming normal body temperature and a relatively low room temperature (20°C), the Carnot efficiency is: (In a warmer environment (27°C) the Carnot efficiency drops to 3.2%)

$$\frac{T_{BODY} - T_{ambient}}{T_{BODY}} = \frac{310\,K - 293\,k}{310\,K} = 5.5\% \quad 2.2$$

The energy that can be recovered using the Carnot engine is in the range of 2.4 W–4.8 W. Taking into account that the total amount of power that is wasted in the form of heat it is around 100 W, this represents a low efficiency.

While a full wetsuit or even a torso body suit is unsuitable for many applications, the neck offers a good location for a tight seal, access to major centres of blood flow, and easy removal by the user. The neck is approximately 1/15 of the surface area of the "core" region (those parts that the body tries to keep warm at all times). As a rough estimate, assuming even heat dissipation over the body, a maximum of 0.20–0.32 W could be recovered conveniently by such a neck braces.

The head may also be a convenient heat source for some applications where protective hoods are already in place. Furthermore, the head is also a very convenient spot for coupling sensory input to the user is some biomedical applications. The surface area of the head is approximately three times the neck area and could provide 0.60–0.96 W with an optimal conversion.

There are different approaches to recover this energy thanks to thermoelectric generators. Low-power thermoelectrically-driven products have appeared. Produced during the 1990s, the Seiko Thermal-wristwatch [51] uses ten thermoelectric modules to generate a microwatt from the small thermal gradient provided by body heat over ambient temperature to activate its mechanical clock movement.

Other company, Applied Digital Solutions, developed a thermoelectric generator called Thermo Life [57] with an area of 0.5 cm^2 and 1.6 mm thickness. It can generate 10 μA at 3 V (6 V open circuit) with only 5°C of temperature difference thank to its dense array of low-temperature thermopiles.

The ThermoLife is designed to power low-current Biosensor electronics working in contact with the skin. These systems typically come with batteries that store extra energy produced during periods of higher temperature differences, so they can work continuously during less efficient ambient temperatures.

The motion of the body is also a feasible power source for harvesting. Several normal movements of the human body can be used as energy sources as is depicted in the body map of Fig. 1.3

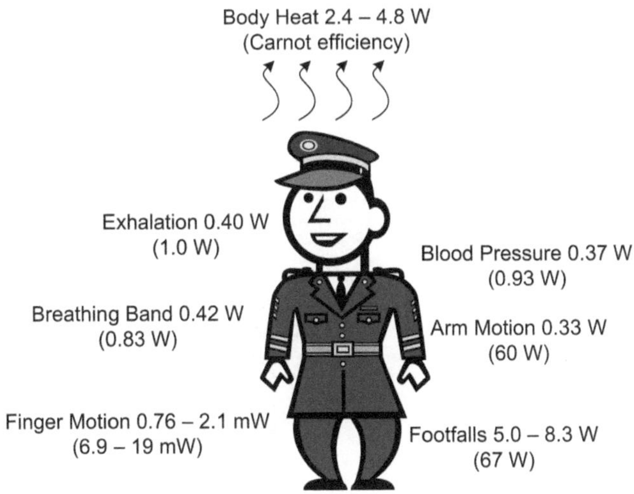

Fig. 1.3 Possible power recovery from body-centered sources. Total power for each action is included in parentheses [21]

Table 1.3 Some examples of active human generation by specific movements (Activity)

Activity	Power generation (W)
Finger (pushing pen)	0.3
Legs (cycling 25 km/h)	100
Hand and arm (freeplay)	21
Hand (aladdinpower)	3.6

Table 1.3 Presents some examples of generated energy with human movements. As can be seen, the amount of power generated depends on the activity and the moved limb

In [58] appears the idea of an energy fitness club where people could reload their portable device while getting in shape. The conception of the pedaling work could be used to supply different consumer electronics, Fig. 1.4 It is estimated that 10 min pedaling on a bicycle could generate 2 W and reload a mobile phone, which is described by the authors as HUMAN POWER.

Other foot-driven generators are not based on bicycles and use a small stationary pedal coupled to an embedded magnetic generator. A perfect example is the "Step charger" manufactured by Nissho Engineering able to generate up to 6 W when the pedal is vigorously pumped.

The Nissho Aladdinpower®, Fig. 1.4, is another product that uses human power. It is a hand-powered device that supplies power to electronic devices. IT is generated enough power to partially recharge a small battery when squeezing it for a minute.

Also, Freeplay [59] is a great example of an enterprise working in this field. This company is solely dedicated to human powered (also called free-powered) products.

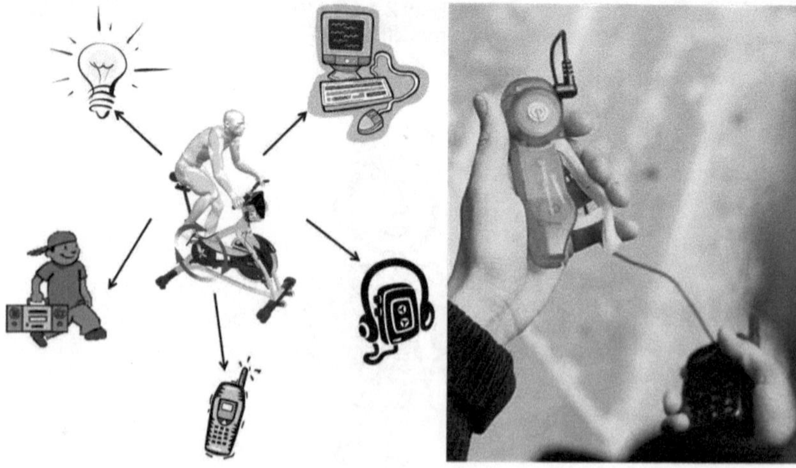

Fig. 1.4 Power recovery from pedaling activity [58] and the Aladdinpower® charger

Fig. 1.5 Examples of Freeplay products. From left to right, the Indigo Self-Sufficient LED Lantern, a rechargeable FlashLigh and the ZipCharger ultra fast charger [59]

Some of their products are flashlights, lanterns, or portable zip-charger based on the movement like those shown in Fig. 1.5.

The walking activity is also a source of energy. The MIT Media Lab. has developed a full system able to harvest parasitic power in shoes employing piezoelectric materials [60, 61], Fig. 1.6. The low-frequency piezoelectric shoe signals are converted into a continuous electrical energy source. The first system consists in harvesting the energy dissipated in bending the ball of the foot by placing a multilaminar PVDF bimorph under the insole. The second one consists in harvesting the foot strike energy with a semi-flexible PZT under the heel. Both devices were excited under a 0.9 Hz walking activity.

Fig. 1.6 A pair of sneakers equipped with piezoelectric power-generating insoles and self-powered transmitter [60]

Industrial applications could also be developed using passive power harvesting. A nice example is the piezoelectric energy conversion at the Tokyo metro station [59]. They expect to obtain over 1,400 kW per day thanks to the walking of the passengers.

On the other hand, human harvesting could be also applied in biomedical and e-health monitoring. Figure 1.7 presents the conception of ubiquitous Body-Sensor network with external and implanted Biosensor powered by human harvesting [61].

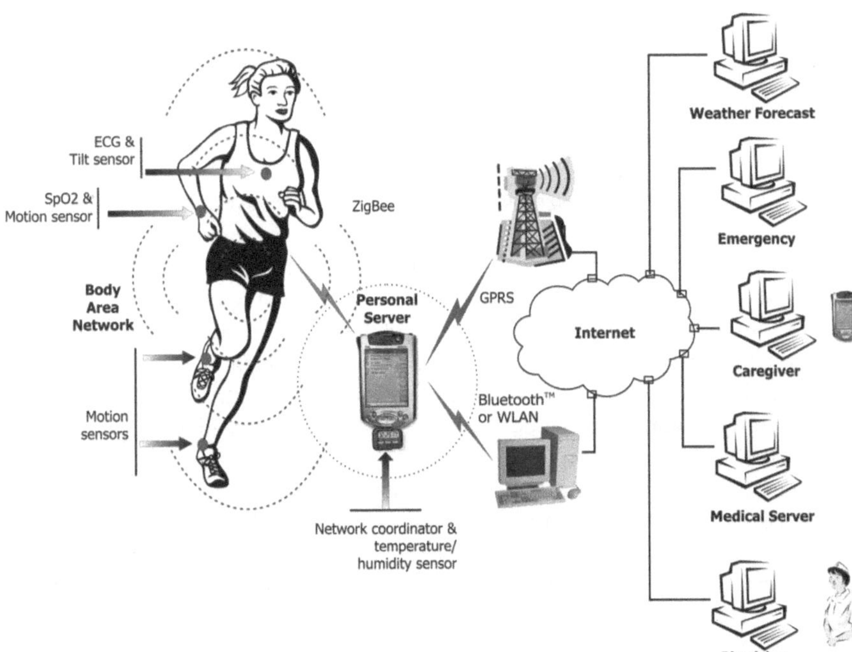

Fig. 1.7 Example of on-body sensor network [61]

Fig. 1.8 (*Right*) Smart T-shirt by Holst/IMEC [64] and (*Left*) RFID tag by Verichip [65]

Some examples of Body-Sensor areas can be found in the literature and from several companies like Toumaz [62]. An interesting one is the smart T-Shirt developed by IMEC [63] and depicted in Fig. 1.8. It is powered by the body heat and used to monitor the cardiac activity.

This IMEC [66] invention demonstrated the integration of a wireless autonomous sensor system in clothes. The system is fully autonomous for its entire life and requires no service – like replacing or recharging the battery – from the user. The shirt with integrated electronics can be washed in a regular washing machine.

The device is powered from a rechargeable battery. The battery is constantly recharged through thermoelectric conversion of the wearer's body heat. The Thermoelectric Generator (TEG) is divided into 14 modules to guarantee user comfort. It occupies less than 1.5% of the shirt area and typically generates a power of 0.8–1 mW @ 1 V at regular sedentary office activity. However, the power increases up to 2.7 mW at 22°C due to forced convection if the user walks indoors

Other alternative to recover energy on the body comes from the Radio Frequency (RF) or Magnetically Induced radiations. Power radiation is a potential power reservoir for on-body application because lots of RF and wireless transmissions are present in the urban environments. However, the development of electronic systems able to collect energy from radiation sources tends to be extremely power-limited and generally requires a large collection area or a close location to the radiating source [10, 36, 12, 67].

Some examples of magnetically induced power harvesting in health applications are the devices developed by VeryChip [65, 67] and Health Link [68]. Health Link utilizes a tiny, passive microchip and a private online database that links the user to the personal health record. It provides emergency room doctors and nurses with the vital medical and emergency contact information, allowing them to treat the human being rapidly, accurately and safely during an emergency. Meanwhile, Verichip has developed an implantable RFID Tag powered by an inductive link through the skin, Fig. 1.8.

1.2 Biosensors

A Biosensor is defined as a "device that uses specific biochemical reactions mediated by isolated enzymes, immunosystems, tissues, organelles or whole cells to detect chemical compounds usually by electrical, thermal or optical signals" [69]. A Biosensor that includes transducers based on integrated circuit microchips are often referred to as Biochip.

In recent years, many types of Biosensors have been developed and used in a wide range of biomedical and other applications. This is a fast-moving field and it is beyond the scope of this work to perform a long dissertation about Biosensors but it is here introduced the enormous importance and potential of these devices.

Basically, a Biosensor could be described as faradic and non-faradaic sensors. In faradaic Biosensors a Redox species is alternately oxidized and reduced by the transfer of an electron to and from the metal electrode; so, a Redox-active species is necessary. On the other hand, in non-faradic sensors no specific or additional reagent is required. Sometimes, a non-faradaic sensor used at a single frequency of operation is designated as capacitive Biosensor.

Thanks to the advances in Biosensors design, some of them are used in the analysis of trace substances in environmental science, pharmaceutical and food industries. However, the continuous monitoring mode is still a challenge for the sensors research.

One of the earliest references to the concept of a Biosensor is from Dr. Leland C. Clark who created many of the early Biosensors in the early 1960s [70]. He used an "enzyme electrode" for measuring glucose concentration with the enzyme Glucose Oxidase (GOD). The success of single analyte sensors was followed by the development of integrated multi-analyte sensors capable of more comprehensive analyses, such as a single instrument for glucose, lactate, and potassium detection.

Technical developments in manufacturing enabled the development of miniaturized integrated Biosensors for the determination of glucose, lactate, and urea in micro samples of undiluted whole blood or plasma. Miniaturization also allowed additional analytical tools to be added to the Biosensor, such as chromatography or capillary electrophoresis.

The newest generation of Biosensors includes miniaturized multi-analyte immunosensor devices with high-throughput capabilities and more than 1,000 individually addressable electrodes per square centimeter. These instruments can detect analytes present in the attomole range [71]. Furthermore, modern fabrication techniques such as ink-jet printing, photolithography and microcontact printing are contributing to the development of more advanced Biosensors that will allow the design of new miniaturized Biosensors with high-density ligands, self-contained lab-on-a-chip capabilities and in the nanoscale range.

Current, conductivity, capacitance variations, and impedance measurements are some of the reported ways to electrically monitor biochemical interactions. Current research in Biosensor technology has been toward developing better transducers to demonstrate superior sensitivity, portability, accuracy and throughput. Parallel, high-throughput monitoring of interactions between biomolecules is playing a significant role in the characterization of various protein functions.

Some interesting examples of health care Biosensors are lactate, urea, cholesterol, uric acid, DNA and inmunosensors. Lactate measurement is helpful in respiratory insufficiencies, shocks, heart failure and metabolic disorder and to monitoring the physical condition of athletes whereas urea estimation is important in monitoring kidney functions and disorders associated with it. Determination of cholesterol is clinically very important because abnormal concentrations of cholesterol are related with hypertension, hyperthyroidism, anemia and coronary artery diseases. Uric acid is one of the major products of purine breakdown in humans and therefore, its determination serves as a market for the detection of disorders associated with altered purine metabolism, notably gout, hyperuricaemia and Lesch–Nyhan Syndrome.

DNA Biosensors have an enormous application in clinical diagnostics for inherited diseases. Immunosensors are based on the specific interactions between antibodies and antigens agents, and are of great interest to analyze microorganisms, viruses, pesticides and industrial pollutants. The variations in the electronic signal to be measured are due to the antibody-antigen (Ab-Ag) interactions.

Other well known example of application is the glucose Biosensor [72–74] based on the electron transfer during the enzymatic reduction of glucose. In the last years, several works have been published regarding the Glucose Continuous Monitoring (GCM) sensor and the conception and implementation of artificial pancreas, like in [75] where it is presented an electro-enzymatic glucose sensor or in [76] where a capacitive based MEMS affinity sensor for continuous glucose monitoring applications is designed.

Rodrigues et al. [77] have developed a new cell-based biochip dedicated to real-time monitoring of transient effluxes of glucose and oxygen with arrays of amperometric microsensors integrated in the inlet and the outlet of a PDMS cell chamber. More advanced and complete sensor is presented in [78] where the design, microfabrication, packaging, surface functionalization and in-vitro testing of a complete electrochemical Cell-on-Chip for continuous amperometric monitoring of glucose is demonstrated performing some cyclic voltammetries, electrical impedance spectroscopy (EIS) and a microscopic examinations.

The use of smart new nanomaterials is of special interest. These materials are sensitive to chemical and biological interactions and they open the door to develop NanoBioSensors able to deal with molecules. They can be integrated into other technologies such as Lab-on-Chip to facilitate molecular diagnostics allowing a wide range of applications like the detection of microorganisms in various samples or in body fluids and detection of tissue pathology such as cancer. Their portability makes them ideal for pathogenesis of cancer (POC) applications.

Some examples related with glucose sensing and NanoBioSensors are presented in [79, 80] where electrodes based on carbon nanotubes and nanoporus of polypyrrole (PPy) respectively are reported to define amperometric glucose monitoring. ZnO nanowires directly connected to the gate of a standard low-threshold MOSFET are also used for Glucose Continuous Monitoring (GCM) as in [81].

All previous examples are focused on using amperometric signals to detect the desired reaction. On the other side, electro-inactive reactions like antibody-antigen affinity binding, where no direct amperometric signal are generated, the changes that occurs in the impedance of the bioactive film have to be measured [82–86]. In [87] is introduced an antiboby-based immunosensor that used Polyaniline nanowire bundles to transducer binding events between pathogens and their target antibodies into variations of the conductance. The variations in the concentration of antigens in the sample are captured as current variation across the electrodes, changing the measured impedance. A model of the variation of the impedance in terms of the concentration of the antigens then is used to analyze the performance of the sensor.

The electronics developed in this work is focused on low-frequency amperometric, and capacitive [88, 89] Biosensors. An amperometric sensor generates a current when a potential is applied on the electrochemical cell, which is proportional to the electrochemical reactions produced in the Bioreceptor. On the other hand, capacitive Biosensor produces a variation on its capacitance when a reaction takes place in the Bioreceptor. In that way, an impedance variation is produced in the electrode-solution interface when an AC signal is applied on the cell [90], which is impedance spectroscopy detection, at a fixed operating frequency.

This means that an amperometric sensors gives the electrochemical information by modulating the current through it, whereas, capacitive sensors gives the information by modifying the sensor impedance.

There are many examples of Biosensors and the interaction with the control and processing electronics. Some hybrid and monolithic integrations can be found in the literature for in-vivo or in-vitro applications like microelectronic pills [91], miniature wireless biopotential recording system [92], multichannel neurotransmitter sensing circuits [93], CMOS sensor interface arrays for DNA detection [94] or CMOS Lab-on-Chip [95].

The novel system presented further on is able to characterize the sensors thanks to a cyclic voltammetry and amperometric measurements in order to detect the variation of concentration of a specific agent. Also, the electronics explore the capability to define an impedance analysis (EIS) through an integrated lock-in amplifier.

1.2.1 Three Electrodes Biosensors

It is important to analyze the definition of the number of necessary electrodes when working with Biosensors. A basic electrode sensor is formed by two electrodes (Reference (R) and Working (W) electrode). The measurement is usually done by applying a voltage signal between them (V_{RW}) when they are inserted in a chemical environment. At the same time, the current across them is measured. The voltage V_{RW} must remain invariable during the measure to achieve a reliable measurement. Unfortunately, the polarization of the electrodes varies due to the current circulation across them, nullifying the measurement. This problem could be solved including an extra electrode (Auxiliary (A) or Counter (C) electrode), which provides the supply current to the working electrode, so no current flow through the reference electrode.

The three electrodes configuration is usually adopted. The three-electrodes are: a) the working electrode (W), which serves as a surface on with the electrochemical reaction takes place; b) the reference electrode (R), used to measure the potential at the W_E, and c) the counter electrode (C), or also called auxiliary electrode (A), has the role to supply the current required for the electrochemical reaction at the W.

The three electrodes topology provides greater stability and reliability than a two electrodes topology. It avoids the polarization of the reference electrode and hence, it is a constant voltage supply between the reference and work electrodes. It is very important that the voltage between the reference and working electrodes (V_{RW}) remains equal for every working conditions being independent of the environment in order to be able to do an accurate measurement and a correct study. A detailed introduction to 3 electrodes sensors is done in Chapter 3.

Several factors contribute to the current generated by the sensor [96], but the size of the working electrode is one of the most important ones. The amount of current generated takes considerable importance in how works with the sensor and the electronics. If the size of the electrodes is decreased, defining microsized electrodes, the current level could decrease to femtoamperes

1.3 Circuits for Three Electrodes Biosensors

Electronic measurement of the biochemical concentrations is essential for diseases diagnosis and study of biological systems. Taking into account that the conception of the system is based on the use of 3 electrodes sensors, the electronics are developed following this premise.

The electronics circuit known as "Potentiostat" is basically in charge to manage the sensor and the techniques used to characterize the sensor like the Electrochemical Impedance Spectroscopy (EIS) or Cyclic Voltammetry (CV)

1.3.1 Potentiostat

It is the key component when working with three electrodes Biosensors. It can be configured in two modes (a) the potentiometric configuration, where a fixed current is applied and the output voltage is measured, or (b) the amperometric configuration, where a fixed voltage is applied and the output current is measured through a current measuring system, for instance, a Transimpedance amplifier (TIA).

The design of a potentiostat varies from discrete to mixed and full-custom ASIC designs. The definition of the adopted solution is given by the size of the sensor and complexity of the system. For electrodes areas greater than 1 mm^2 just electronics based on low-cost surface mount components can be adopted but, for smaller and multiplexed solutions, full-custom ASIC solutions appears as a valid approach. Furthermore, the conception of an ASIC interface near the electrodes is motivated when very low current levels are derived from the sensors. Other key aspects to use an ASIC solution near the electrodes are, a part of the degree of miniaturization: a) EMI can be reduced; b) External disturbances like vibrations, moisture, sources of electrical noise, etc… are avoided.

Some references are described in Choi Myung-suk et al [89] in their work “Implantable Bio system design for displacement measurement of living life” or in the work by K. Kitamori [90], where he described micro and nano chemical sensors on-a-chip. Other interesting Biosensors are the piezoelectric immunosensors, like the one developed for the rapid diagnosis of M. tuberculosis by Eric Carnes et al [92].

Several references related the use of ASIC designs and Biosensors are presented in [97, 98] and [99], which introduces a low power readout circuit with a potentiostat amplifier for amperometric chemical sensors in Glucose Meter Application.

A basic potentiostat is divided in two parts: 1) the control and 2) the measure (TIA) and post-amplification circuitry. The control circuitry is used to maintain a constant voltage across the R and W electrodes. On the other hand, the TIA stage is in charge of detecting the current generated in the reaction.

One of the most used potentiostat architecture is depicted on Fig. 1.9 where it is also depicted one of the most usual electrical model for electrochemical cells [100], the Randles RC model. This simple and compact architecture is interesting for ASIC implementation [101–103] and to develop applications with multi-working sensors because the current is measured independently for each sensor with a dedicated read-out circuit.

This structure is based on four operational amplifiers (Opamp), and two resistors. OP4 is the Transimpedance (TIA) amplifier, which defines the virtual ground voltage of the W electrode and provides current-to-voltage conversion such that,

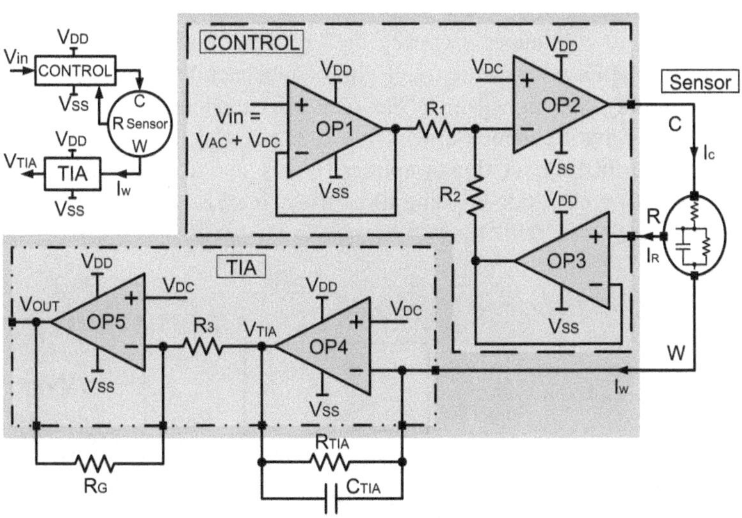

Fig. 1.9 Schematic view of the Biopotentiostat amplifier with the electrochemical electrical model

$$V_{TIA} = -I_W R_{TIA} \; 1.1$$

where I_W is the current through the cell and R_{TIA} the gain resistance defined in the Transimpedance amplifier.

OP3 is used to ensure minimal current flows through the reference electrode and to sense the voltage difference between the R and W electrodes. This difference is compared with the desired V_{in} by OP2 in order to drive properly the counter (C) electrode and to maintain controlled the V_{RW} voltage.

The architecture in Fig. 1.9 is the selected one to design the implantable electronics and it is explained in detail in Chapter 3.

The previous potentiostat architecture uses a classical Transimpedance amplifier configuration to detect the current. In order to solve the frequency bandwidth limitations associated with the standard TIA, other options could be studied for the analogue readout of the signal like the ones presented in Fig. 1.10. This converter is based on an integrator-differentiator scheme to combine low-noise and wide bandwidth with two simultaneous outputs.

A high speed output (V_{ac}) amplifies the input current by the capacitive ratio C_d/C_i and converts it into voltage by the resistor R_t. Differently from a standard Transimpedance amplifier, where the $I-V$ conversion is set by a feedback resistor, whose value is necessarily large also for having low noise at the expense of a dramatic drop of the bandwidth, in this solution the current amplification is set by means of capacitors, noiseless components, and allows a small value of R_t, without affecting the overall noise performance.

The bandwidth is largely increased, and is set by the integrator loop-gain to the value $BW = GBP \cdot (C_i/(C_i + C_c))$ where GBP is the gain-bandwidth product of Opamp1 and C_s is the input capacitance on the input node.

Since a DC input current would saturate the integrator stage, an active feedback has been added to continuously convey the DC input current into a proper sink (resistor R_{dc}). The Low Pass Filter (LPF) in the feedback branch must be designed as an active filter with high gain in the very low frequencies and a very strong attenuation for higher frequencies.

The use of an Instrumentation Amplifier (IA) is also a widely used alternative if just one working electrode is going to be used. It adds an extra resistor (R_a) in

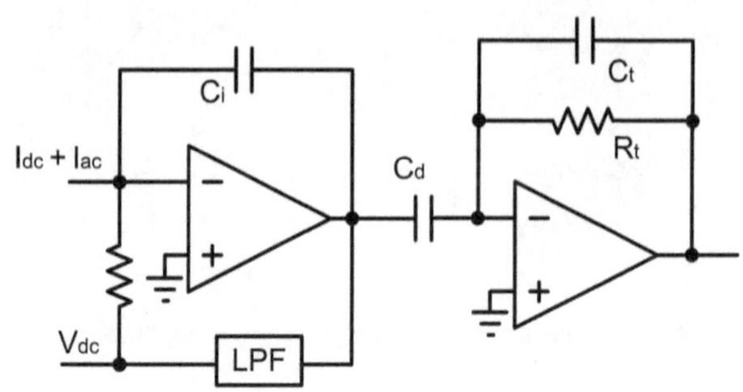

Fig. 1.10 Alternative transimpedance amplifier configuration

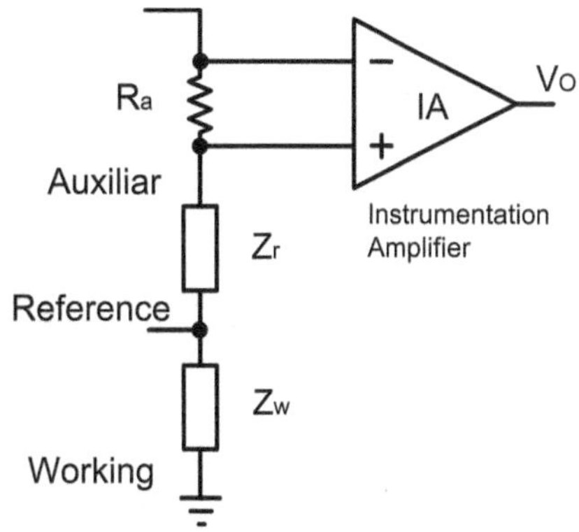

Fig. 1.11 Instrumentation amplifier stage

series with the Counter electrode to detect the current, Fig. 1.11. In that way, the Working electrode is physically connected to ground and not to a virtual ground as in the previous Transimpedance solution.

1.3.2 Discrete Potentiostat Approach

The architecture in Fig. 1.9 has been selected for developing the CMOS implantable instrumentation, although the structure has been previously validated and tested through a discrete implementation before the IC design. A first discrete prototype has been implemented for O_2 monitoring in environmental applications. The derived electronics were also used to validate the merge of these electronics with the electrodes in the conception of an encapsulated device.

This development combines a fluidic channel device integrated with the electrodes and the electronic system for oxygen monitoring in water to form a disposable device.

Dissolved oxygen concentration (DO) is an important index of water quality and the ability to measure the oxygen concentration and temperature at different positions and depths would be an important attribute to environmental analysis. The decrease of oxygen concentration in water is a clear indication of water pollution, which is one of the main concerns of the Water Framework directive in the European Union [104]. Moreover, in other topic, the detection of O2 concentration in blood plasma has several interesting applications including the oxygen monitoring during open-heart surgery [105].

This discrete system is based on sensors and electronics forming one module as is depicted in Fig. 1.12. The discrete PCB potentiostat amplifier is coated with PDMS and fully characterized through several Cyclic Voltammetries. The PCB was tested before and after its introduction into water

The custom electronics, which includes a small printed circuit board (PCB), has a size of 31 mm × 21 mm. The power consumption of this first implementation is

Fig. 1.12 PCB capture coated with PDMS

around 350 mW using the Texas Instrument commercial amplifiers OPA 656 and OPA 657, which acts as a Transimpedance amplifier. These amplifiers were selected for their good characteristics in terms of high input impedance, low bias and offset input currents and voltages.

The obtained results are compared with the ones obtained with the commercial instrument CH 1232A (CH Instruments ®). Figure 1.13 shows some voltammetries obtained analyzing tap water with high concentration (Region A) and low

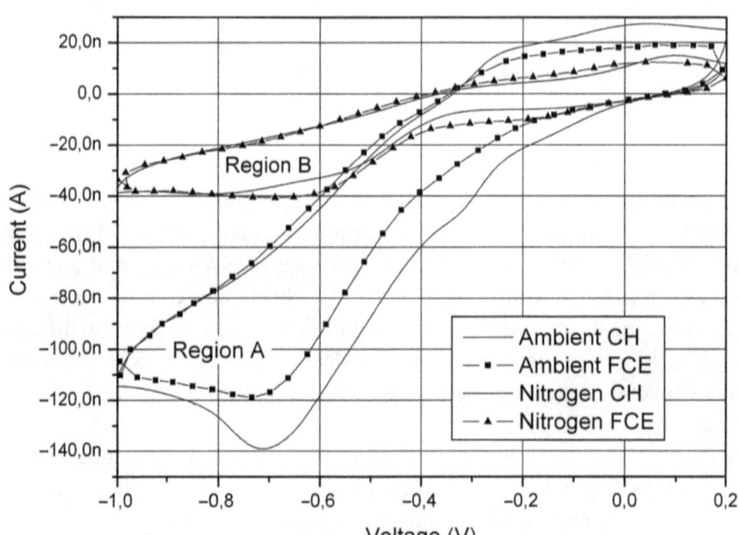

Fig. 1.13 Voltammograms obtained with the CH instrument and with the PCB for TAP water in ambient conditions and with nitrogen

concentration of O_2 (Region B). The more negative current peak around -1 V shown in Region A comes from a sample of water that has been exposed to ambient air.

On the other hand, the less negative peak on Region B is obtained for a water sample bubbled in nitrogen to remove the oxygen. As can be observed, the performance of the custom potentiostat is demonstrated obtaining measurements very close to the commercial one; hence, it is possible to detect several concentrations of O_2 with the discrete full-custom design.

The development of this discrete prototype has been used to acquire background knowledge and to observe problems when working with electrochemical sensors before proceeding with the design of implantable system presented in Chapter 4.

1.3.3 Lock-In Amplifier and Complex Impedance Detection

Regarding the electrochemical characterization of the Biosensors, the signal processing of the obtained measurements by the electrochemical impedance spectroscopy technique, defines a circuitry that has the role to obtain the real and imaginary components of the measurement of the Electrochemical Impedance.

EIS [106] is an AC method that describes the response of an electrochemical cell to a small amplitude sinusoidal voltage signal as a function of frequency. The resulting current sine wave differs in time (phase shift) with respect to the perturbing (voltage) wave, and the ratio V(t) / I(t) is defined as the impedance (Z), and accounts for the combined opposition of all the components within the electrochemical cell (resistors, capacitors, inductors) to the flow of electrons.

In an electrochemical cell, electrode kinetics, Redox reactions, diffusion phenomena and molecular interactions at the electrode surface can be considered analogous to the above components that impede the flow of electrons in an ac circuit. The simplest electrical model is based on an equivalent RC circuit so called a Randles circuit [100].

Based on the nature of the measured signal there are two main approaches:

- the Capacitive sensors where the surface of the electrode is completely covered by a dielectric layer and the whole electrode assembly behaves as an insulator. The variation of the capacitance is measured, in frequency ranges up to 100 kHz.
- and the faradaic sensors. In this case the surface of the electrode, which is partially or wholly covered by a non-insulating or insulating layer, is able to catalyse a Redox activity that exists in the solution. In this case, the measured parameter is the charge transfer resistance (the real component of impedance at low frequency values, typically 0.1–1.0 Hz).

In order to proceed with the signal processing there are mainly two approaches: (1) the Fast Fourier Transform (FFT), and (2) the Frequency Response Analyzer (FRA). In the case of the FFT, a pulse or a step is applied to the sample because it contains wide frequency content. Then, the response of the sample is digitized and processed in a digital processor and the different frequency components are obtained for their analysis using the FFT algorithm. Also, other possibility that could be followed is the logarithmic sampling in the DFFT calculus, reducing the data that must be required in the process.

A simpler solution is based on the FRA approach. In this case a sine and cosine signals are adopted and using two multipliers and a filter stage the real and imaginary components of the response are obtained. This measurement must be done for each frequency to obtain the full EIS characterization of the Biosensor. Working with just one sensor and in terms of the size of the final product, the FFT option could be adopted, because the response for several frequencies is obtained.

The FRA solution is more oriented to multi-sensor approaches but, also in the case of single sensors, it is a nice option in terms of complexity and speed for not too low frequencies. This FRA or Lock-In approach is more feasible and it is an interesting solution when just the impedance must be obtained at a fixed frequency of operation.

The perturbation signal provided by the instrumentation system – following the FRA approach – generally use a sinus wave as the input voltage. In this work a low-voltage integrated CMOS lock-in amplifier, Chapter 3, is analyzed as a method to measure the impedance value of the sensor. Typically, a Lock-In amplifier is used to increasing the Signal-to-Noise ratio of the system by synchronizing the output signal to the input signal at a known frequency; i.e., the frequency where the sensor has the best sensitivity response filtering the other frequencies and noise.

Some examples of FRA and Lock-In implementations can be found in the literature. In [107] a fully CMOS lock-in amplifier is proposed in the field of gas sensors that works at a fixed operating frequency. In [108] is derived a more complex lock-in amplifier for electrical bio-impedance analysis for implantable medical devices. There, the pulse signals used for the synchronous demodulation channels are modified by specific dead times so, the error in the magnitude and phase is reduced.

Reference [100] describes this technique as a very powerful tool for the analysis of complex electrochemical systems and how can provide a valuable and complete description of the electrode process. Following this idea, a theoretical analysis and design of a CMOS Lock-In amplifier for bio Impedance detection is carried out in Chapter 3.

1.4 Contribution of this Book

The main objective of this book is the conception of a generic Self-Powered CMOS Front-End Low-Voltage Low-Power integrated device that would be the main part of a subcutaneous device for biomedical applications [109–116]. At the end of this work, a prototype device would be implemented able to manage the information coming from a 3 electrodes amperometric sensor.

The devices, designed as a subcutaneous implant (small dimensions), will validate the conception of Implantable Biomedical event detector able to discriminate several specific concentration points of the analyzed substance for that reason the design and conception of the CMOS based implantable architecture has been divided into three parts:

1. the design of a powering system for the electronics,
2. the development of the integrated sensor instrumentation, post-processing and communications,
3. and the design and validation of the implantable detection prototype.

The powering scheme is an important aspect since the device is located under the human skin and the use and maintenance of batteries or conventional energy sources is a drawback. Moreover, to develop a self powered subcutaneous device, an appropriate alternative is to design a system that recovers energy from the environment.

Following this alternative, an specific novelty IC circuit called MHPC (Multi Harvesting Power Chip) has been designed to solve the power problem regarding the subcutaneous ambient of application. It is able to recollect energy from several ambient sources (vibrations, light, EM waves) in the range of mW and to power some low power applications focused on on-body applications. However the MHPC is originally oriented to on-body applications, it could be also used in different type of applications such as main power source in distributed sensor networks.

Basically, the designed MHPC power management is designed to

- assure a high efficiency low-power energy conversion,
- combine several energy sources with the same power management unit,
- low-power consumption,
- and occupy a small size and package.

The second part develops a "BioChip" with the necessary instrumentation and post-processing circuits to work with the three electrodes sensors. This IC integrates the potentiostat to drive and obtain the information from the sensor and the post-processing electronics to detect several values of concentrations.

Low-power circuits have been implemented to design the instrumentation and to avoid power consumption higher than 1.5 mW. In that way, it is possible to reduce damages under the human skin due hot spots. Furthermore, the theoretical and simulated design of a lock-in amplifier to detect complex impedance in Biosensors is also described as a part of the post-processing.

Finally, an ASIC based Self-Powered implantable device prototype is fabricated to validate the novel generic architecture for In-Vivo even detection with amperometric Biosensors. The prototype is based on two IC's sharing the same PCB substrate.

The MHPC chip will harvest the energy from an inductive link through the human skin and will generate the necessary on-chip voltages to drive the BioChip circuits. On the other side, the BioChip will drive the sensor and will deal with the post-processing to detect the desired concentrations. Moreover, the inductive link is also used to transmit the detected concentrations to an external receiver in the other side of the skin.

The design of the MHPC can be also oriented to work as a main power source to develop a Global (implantable) Application where the external device (or receptor) is also powered through harvesting sources [117, 118]. Figure 4.14 shows the conception of a Global Implantable Application where all the involved electronics (external and internal) are fully self-powered using the MHCP. Furthermore, in the future, the MHPC could use newer implantable power sources (Mecatronics, Fig. 4.14) to drive the subcutaneous device.

In short, this book presents novel low-power architecture for portable subcutaneous detection device powered through the energy harvesting concept. Thereby, the development of this kind of implantable and portable detectors represent a step forward in the biomedical diagnosis and detection devices evolution as it is illustrated in Fig. 1.15. Moreover, this work is the perfect link between the development of biomedical subcutaneous devices and the use of alternative energies, two of the major emerging research fields, and demonstrates that the development of complex biomedical detection systems could be completely self-powered.

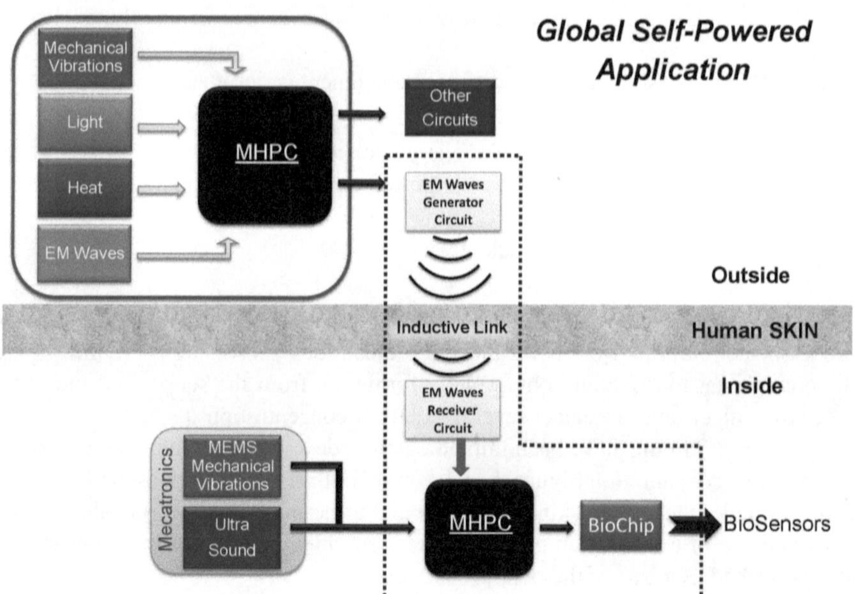

Fig. 1.14 Conception of the proposed global implantable subcutaneous application where implanted and external electronics are self-powered (autonomous). The *dotted line* represents the work covered in this book

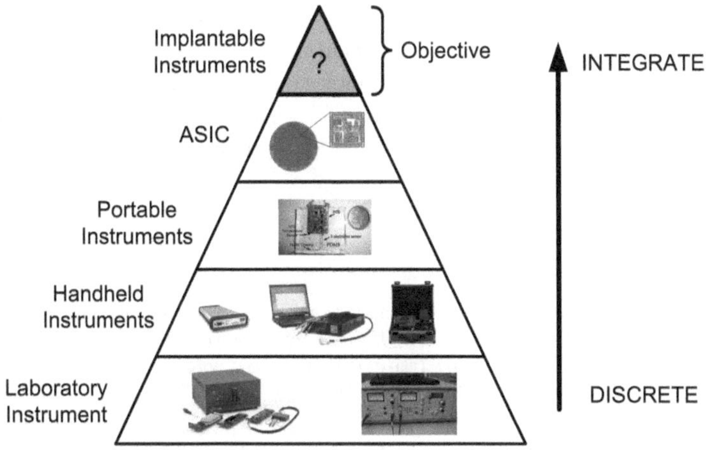

Fig. 1.15 Pyramid with the evolution of biomedical diagnosis equipment

1.5 Outline of the Book

This book is divided in three main chapters. The first one, Chapter 2, develops the Multi Harvesting Power Chip (MHPC). On it, all circuits used to recollect energy from several ambient sources are theoretically and experimentally described and validated. Finally, the whole chip is tested working with all circuits collecting energy form more than one source at the same time.

On the other hand, Chapter 3 is focused on the development of a BioChip with the integrated instrumentation to work with three electrodes amperometric Biosensors. First of all, it is introduced the conception of three electrodes Biosensors and how they work. Voltammetry, EIS and amperometric techniques are also introduced to the reader.

The potentiostat architecture is clearly explained and experimentally tested using a real substance. The obtained results with the full-custom approach are compared with the ones obtained using a commercial potentiostat. In that way, the correct operation of the designed instrumentation is fully validated.

Furthermore, this chapter explains the conception of a Lock-In amplifier circuit used to detect the real and imaginary parts of the complex impedance coming from the Biosensor. This circuit is theoretically explained and some simulated results are shown. It is also introduced on this chapter the conception of Biotelemetry or how to transmit information from the subcutaneous device to the external reader. Then, the implemented protocol in this work is detailed.

Chapter 4 finally describes the design and conception of the Self-Powered CMOS Front-End architecture for Biomedical Subcutaneous Devices. All circuits are presented in detail as well as the powering through an Inductive link. The power and communication antenna as well as the connections between the MHCP IC, the BioChip IC and the sensor are also explained.

The results obtained with the final capsule prototype with a size less than 4.5 cm × 2.5 cm are shown and commented in detail. It is analyzed the problems regarding the misalignment between the internal and external antennas. Finally, the prototype has been validated as a detector. To conclude with the book, Chapter 5 presents the conclusions and possible future options that may arise from this work.

References

1. C. Sauer, N. Thakor, Power harvesting and telemetry in CMOS for implanted devices. IEEE Trans. on Circuits and Syst. **52**(12), 2605–2613 (Dec 2005)
2. Y. Li, J. Liu, A 13.56 MHz RFID Transponder front-end with merged load modulation and voltage doubler-clamping rectifier circuits. IEEE Int. Sym. on Circuits and Syst. 5095–5098 (2005)
3. K. Myny, et al, An inductively coupled 64b organic RFID tag Operating at 13,56 MHz with a data rate of 787b/s. IEEE Int. Solid-State Circuits Conf. 290–614 (2008)
4. A. Gore, et al., A multi-channel femtoampere-sensitivity conductometric array for biosensing applications. 28th IEEE Eng. in Medicine and Biology Science Conference, pp. 6489–6492, 2006
5. M. R. Haider, et al., A low-power processing unit for in vivo monitoring and transmission of sensor signals. Sensors & Transducers J. **84**(10) 1625–1632 (Oct 2007)
6. J. Colomer-Farrarons; P. Miribel-Català; J. Samitier; M. Arundell; I. Rodríguez, Design of a miniaturized electrochemical instrument for in-situ O2 monitoring. (Proceedings Paper Spie'09), VLSI Circuits and Systems IV, Microtechnologies for the New Millennium, Volumen 7363 (2009)
7. http://www.solicore.com/flexion-batteries.asp
8. http://www.tesa.com/industry/health_markets/transdermal_patches/transdermal-patches-and-oral-dissolvable-films,850384,1.html
9. Transdermal Drug Delivery Promises To Eliminate Needles. http://www.popularmechanics.com/science/health_medicine/1281106.html
10. C.M. Zierhofer, E.S. Hachmair, Geometric approach for coupling enhancement of magnetically coupled coils. IEEE Trans. Biomed. Eng. **43**, 708–714 (1996)
11. M. Sawan, H. Yamu, J. Coulombe, Wireless smart implants dedicated to multichannel monitoring and microstimulation. IEEE Circuits Syst. Mag. **5**, 21–39 (2005)
12. C. Sauer, M. Stanacevic, G. Cauwenberhs, N. Thakor, Power harvesting and telemetry in CMOS for implanted devices. IEEE Trans. Circuits Syst. **52**(12), 2605–2613 (Dec 2005)
13. Y. Li, J. Liu, A 13.56 MHz RFID transponder front-end with merged load modulation and voltage doubler-clamping rectifier circuits. IEEE International Symposium on Circuits and Systems, 5095–5098 2005
14. K. Myny, S. Van Winckel, S. Steudel, P. Vicca, S. De Jonge, M.J. Beenhakkers, C.W Sele, N.A.J.M. van Aerle, G.H.Gelink, J. Genoe, P. Heremans, An inductibely-coupled 64b organic RFID tag Operating at 13,56 MHz with a data rate of 787b/s. IEEE International Solid-State Circuits Conference, pp. 290–614 (2008).
15. O. Meirik, Implantable contraceptives for women. Contraception. **65**(1), 1–2 (2002)
16. O. Meirik et al, Implantable contraceptives for women. Hum Reprod Update. **9**(1), 49–59 (2003)
17. S. Alepuz, S. Busquets-Monge, J. Bordonau, J. Gago, D. Gonzalez, J. Balcells, Interfacing renewable energy sources to the utility grid using a three-level inverter. IEEE Trans. Ind. Electron. **53**(5), 1504–1511 (Oct 2006)
18. J.M. Carrasco, L.G. Franquelo, J.T. Bialasiewicz, E. Galvan, R.C. PortilloGuisado, M.A.M. Prats, J.I. Leon, N. Moreno-Alfonso, Power-Electronic systems for the grid integration of renewable energy sources: A survey. IEEE Trans. Ind. Electron. **53**(4), 1002–1016 (June 2006)

19. J. Schonberger, R. Duke, S.D. Round, DC-bus signaling: A distributed control strategy for a hybrid renewable nanogrid. IEEE Trans. Ind. Electron. **53**(5), 1453–1460 (Oct. 2006)
20. L. Collins, Harvest for the world. IEEE Power Eng. **20**(1), 34–37 (Feb–March 2006)
21. J.A. Paradiso, T. Starner, Energy scavenging for mobile and wireless electronics, IEEE Pervasive Comput. **4**(1), 18–27 (Jan–March 2005)
22. E.M. Yeatman, Energy scavenging for wireless sensor nodes, in Proceedings. of the 2nd International Workshop on Advances in Sensors and Interface, 1–4 (2007)
23. S. Roundy, D. Steingart, L. Frechette, P. Wright, J. Rabaey, Power sources for wireless sensors networks, in Proceedings. of the 1st European Workshop on Wireless Sensors Networks, 1–17 (Jan.2004)
24. D. Niyato, E. Hossain, M.M. Rashid, V.K. Bhargava, Wireless sensor networks with energy harvesting technologies: A game-theoretic approach to optimal energy management. IEEE Wirel. Commun. **14**(4), 90–96 (Aug 2007)
25. N.S. Shenck, J.A. Paradiso, Energy scavenging with shoe-mounted piezoelectrics. IEEE Micro. **21**(3), 30–42 (May–June 2001)
26. T. Starner, J.A. Paradiso, *Human-generated power for mobile electronics*, ed. by C. Piguet. Low-Power Electronics Design, (CRC Press, 2004), Chapter 45, pp. 1–35
27. M.S.M. Soliman, E.F. El-Saadany, R.R. Manssur, Electromagnetic MEMS Based Micro-Power Generator, in Proceedings of the IEEE International Symposium on Industrial Electronics, Vol.4 (Jul 2006), pp. 2747–2753
28. X. Cao, W. Chiang, Y. King, Y. Lee, Energy harvesting circuit with feedforward and feedback DC–DC PWM boost converter for vibration power generator system. IEEE Trans. Power Electron. **22**(2) 679–685, (March, 2007)
29. S. Meninger, J.O. Mur-Miranda, R. Amirtharajah, A. Chandrakasan, J.H. Lang, Vibration-to-electric energy conversion. IEEE Trans. Very Large Scale Integration (VLSI) Systems. **9**(1), 64–76 (Feb.2001)
30. N. Ben Amor, O. Kanoun, Investigation to the Use of Vibration Energy for Supply of Hearing Aids, in Proceedings of the IEEE Instrumentation and Measurement Technology Conference, (May 2007), pp. 1–6
31. Y. Ammar, A. Buhrig, M. Marzencky, B. Charlot, S. Basour, K. Matou, M. Renaudin,Wireless sensor network node with asynchronous architecture and vibration harvesting micro power generator, in Proceedings of the SOC-EUSAI Conference, (Oct 2005), pp. 287–292
32. E.K. Reilly, E. Carleton, P.K. Wright, Thin film piezoelectric energy scavenging systems for long term medical monitoring, in Proceedings of the IEEE International Workshop on Wearable and Implantable Body Sensor Networks, (2006), p. 4
33. D. Puccinelli, M. Haenggi, Wireless sensor networks: Applications and challenges of ubiquitous sensing. IEEE Circuits Syst. Mag. **3**(3), 19–29 (2005)
34. F. Kocer, P.M. Walsh, and M.P. Flynn, *Wireless, Remotely Powered Telemetry in 0.25 μm CMOS*, Radio Frequency Integrated Circuits Symposium, (IEEE Press, FortWorth, Texas, 2004), 339–342
35. N. Cho et al., A 8-μW, 0.3 mm2 RF-Powered Transponder with Temperature Sensor for Wireless Environmental Monitoring, in Proceedings IEEE International Symposium on Circuits and Systems, (May 2005), pp. 4763–4766
36. S.J. Miller-Smith, New Chip Can Read Your Pet's Temperature, Darwin Veterinary Center. www.darwinvets.plus.com/topical/biothermo.htm
37. M. Ferrari, V. Ferrari, D. Marioli, A. Taroni, Modeling, fabrication and performance measurements of a piezoelectric energy converter for power harvesting in autonomous microsystems. IEEE Trans. Instrum. Meas. **55**(6) 2096–2101 (Dec 2006)
38. C.B. Williams, R.B. Yates, Analysis of a micro-electric generator for microsystems. Sens. Actuators A. **52**, 8–11 (1996)
39. M. El-hami, P. Glynne-Jones, N.M. White, M. Hill, S. Beeby, E. James, A.D. Brown, and J.N. Ross, Design and fabrication of a new vibration-based electromechanical power generator, Sens. Actuators A. **92** 335–342 (2001)

40. T. Starner, Human-powered wearable computing. IBM Syst. J. **35** 618–629 (1996)
41. http://www.microsoft.com/presspass/ofnote/11-02worldin2003.mspx
42. J. Brufau, M. Puig, Piezoelectric energy harvesting improvement with complex conjugate impedance matching. J. Intell. Mater. Syst.Struct. 2009 pp. 597–608, (Sept 2008), DOI: 10.1177/1045389X08096051
43. M. Griot, Fundamentals of Vibration Isolation, http://www.mellesgriot.com
44. J.A. Paradiso, Systems for human-powered mobile computing, Proceedings of the 43rd annual Design Automation Conference Annual ACM IEEE Design Automation Conference, San Francisco, CA, SESSION: Session 37: Special session: beyond low-power design: environmental energy harvesting, 645–650 (2006)
45. S. Roundy, P.K. Wright, and K.S.J. Pister, Micro-electrostatic vibration-to-electricity converters proceedings of IMECE'02 2002 ASME international mechanical engineering congress & exposition New Orleans, Louisiana 17–22 (2002)
46. J.A. Paradiso, T. Starner, Energy scavenging for mobile and wireless electronics. IEEE Pervasive Comput. **4**, 18–27 (2005)
47. http://www.nanosolar.com
48. http://www.konarka.com
49. J. Krikke, Sunrise for energy harvesting products, published by the IEEE CS and IEEE ComSoc 1536-1268/05/$20.00 © 2005 IEEE
50. M. Stordeur, I. Stark, Low Power Thermoelectric Generator – Self-Sufficient Energy Supply for Micro Systems, in Proceedings ICT'97 16th International Conferrence Thermoelectrics, (1997), pp.575–577
51. http://www.seikowatches.com
52. http://www.:appliedsolutions.com
53. S. Roundy, P.K. Wright, J.M. Rabaey, *Energy Scavenging for Wireless Sensor Networks* (Kluwer Academic Publishers, Boston MA)
54. http://www.witricity.com
55. "Wireless Power Demonstrated", Retrieved 12-09-2008 http://thefutureofthings.com/pod/250/wireless-power-demonstrated.html
56. Wireless electricity could power consumer, industrial electronics. MIT News. 2006-11-14. http://web.mit.edu/newsoffice/2006/wireless.html
57. Thermo Life Energy Co. http://www.poweredbythermolife.com/
58. T. Kazazian, A.J. Jansen, Eco-desing and human-powered products, proceedings of the Electronics Goes Green, (2004), pp. 6–10
59. FreePlay, To make energy available to everybody all of the time, http://www.freeplayenergy.com/about
60. N.S. Shenck, J. Paradiso, Energy scavenging with shoe-mounted piezoelectrics. IEEE Micro. **21**(3), (May–June 2001) 30–42
61. WHMS – Wearable Health Monitoring Systems, Electrical and Computer Engineering, The University of Alabama in Huntsville, http://www.ece.uah.edu/~jovanov/whrms/
62. Toumaz http://www.toumaz.com/
63. C. Van Hoof, V Leonov, R.J.M Vullers, Thermoelectric and Hybrid Generators in Wearable Devices and Clothes, 6th International Workshop on Wearable and Implantable Body Sensor Networks, 2009. BSN 2009. 3–5 June 2009 195–200
64. Wearable Electronic Science. http://www.fibre2fashion.com/news/nanotechnology-news/newsdetails.aspx?news_id=70079
65. http://www.verichipcorp.com/
66. IMEC http://www2.imec.be
67. Health Care News. "Microsoft Partners With Implantable RFID Chip Maker VeriChip" http://www.rfidjournal.com/article/articleview/4477/1/1/
68. http://www.healthlinkinfo.com/

69. IUPAC Compendium of Chemical Terminology, *International Union of Pure and Applied Chemistry: Research Triangle Park, NC, USA* 2nd edn. (1997, 1992)
70. L.C. Clark Jr, C. Lyons, Electrode systems for continuous monitoring in cardiorascular surgery NY, Acad. Sci. **102**, 29–45 (1962)
71. K. Dill, Biosens. Bioelectron. **20**, 736–742 (2004)
72. M. Hiller, C. Kranz, J. Huber, P. Bauerle, W. Schuhmann, Amperometric biosensors produced by immobilization of redox enzymes at polythiophene-modified electrode surfaces. Adv. Mater. **8**, 219–222 (1996)
73. A. Kros, W.F.M. Van Hovell, N.A.J.M. Sommerdijk, R.J.M. Nolte, Poly(3, 4-thylenedioxythiophene)-based glucose biosensors. Adv. Mater. **13**,1555–1557 (2001)
74. P.A. Fiorito, S.I.C. De Torresi, Glucose amperometric biosensor based on the co-immobilization of glucose oxidase (Gox) and ferrocene in poly(pyrrole) generated from ethanol/water mixtures. J. Braz. Chem. Soc. **12**, 729–733 (2001)
75. J. Patel, B. Kaminska, B. Gray, B. Gates,Electro-Enzymatic Glucose Sensor Using Hybrid Polymer Fabrication Process Electronics, Circuits and Systems, 2007. ICECS 2007. in Proceedings of the 14th IEEE International Conference on 11–14 Dec. (2007) pp. 403–406
76. X. Huang, S. Li, J. Schultz, Q. Wang, Q. Lin, A capacitively based MEMS affinity glucose sensor", Proceedings of the International Solid-State Sensors, Actuators and Microsystems Conference, TRANSDUCERS 2009, (21–25 June 2009), pp. 1457–1460
77. N.P, Rodrigues, H. Kimura, Y Sakai, T. Fujii, Cell-based Microfluidic biochip for electrochemical real-time monitoring of glucose and oxygen Solid-State Sensors, Actuators and Microsystems Conference. (10–14 June 2007) pp. 843–846
78. A. Rahman, G. Justin, A. Guiseppi-Wilson, A. Guiseppi-Elie, Fabrication and packaging of a dual sensing electrochemical biotransducer for glucose and lactate useful in intramuscular physiologic status monitoring Sens. J. IEEE **9** (12),(Dec 2009) 1856–1863
79. J. Xie; S. Wang; L. Aryasomayajula, V.K. Varadan, Material and electrochemical studies of platinum nanoparticle-coated carbon nanotubes for biosensing. Nanotechnology, 2007. IEEE-NANO 2007. 7th IEEE Conference, 2–5 Aug 2007, pp. 1077–1080
80. E.M.I. Ekanayake, D. Preethichandra, K. Kaneto, Fabrication and characterization of nanostructured conducting polymer electrodes for glucose biosensor applications. Industrial and information systems, 2007. ICIIS 2007. International conference, 9–11 Aug 2007, pp. 63–66
81. U. Ali, S.M. Nur, M. Willander, B. Danielsson, Glucose detection with a commercial MOSFET using a ZnO nanowires extended gate. IEEE Trans. Nanotechnol. **8**(6), 678–683 (Nov 2009)
82. Bender, Sadik, Direct electrochemical immunosensor for polychlorinated biphenyls. Environ. Sci. Technol. **32**, 788–797 (1998)
83. G. Farace, G. Lillie, T. Hianik, P. Payne and P. Vadgama, Reagentless biosensing using electrochemical impedance spectroscopy. Bioelectrochemistry **55**,1–3 (2002)
84. E. Katz, I. Willner, Probing biomolecular interactions at conductive and semiconductive surfaces by impedance spectroscopy: routes to impedimetric immunosensors, DNA-sensors, and enzyme biosensors. Electroanalysis **15**, 913–947 (2003)
85. D. Laureyn, P. Nelis, K. Van Gerwen, L. Baert, R. Hermans, J.J. Magnee, G. Maes, Nanoscaled interdigitated titanium electrodes for impedimetric biosensing. Sens. Actuators B-Chem. **68**, 360–370 (2000)
86. Lillie, P. Payne, P. Vadgama, Electrochemical impedance spectroscopy as a platform for reagentless bioaffinity sensing. Sens. Actuators B-Chem. **78**, 249–256 (2001)
87. Y. Liu, S. Chakrabartty, E.C Alocilja, Fundamental building blocks for molecular biowire based forward error-correcting biosensors. Nanotechnology **18** 424-017 (6pp). (2007)
88. J. Colomer-Farrarons, P. Miribel-Català, A. Saiz-Vela, J. Samitier, in *Proceedings. of the 16th IEEE International Conference on Very Large Scale Integration VLSI – SOC*, A 50 μW low-voltage CMOS Biopotentiostat for low frequency Capacitive Biosensor. (Rodhas, Greece, 2008)

89. A. Gore, S. Chakrabartty, S. Pal, E. Alocilja, A multi-channel femtoampere-sensitivity conductometric array for biosensing applications, Proceedings of the Annaual InternationalConference on IEEE Engineeringin Medicine and Biological Society. 2006
90. S.V. Dzyadevych et al. Electrochemical enzyme biosensors (2006), ISBN: 966-02-4200-X
91. E.A. Johannessen, L. Wang, L. Cui, T. Tang, A. Astaras, M. Ahmadian, J.M. Cooper, Implementation of multichannel sensors for remote biomedical measurements in a microsystems format. IEEE Trans. Biomed. Eng. (2004)
92. P. Mohseni, K. Najafi, S.J. Eliades, W. Xiaoqin, Wireless multichannel biopotential recording using an integrated FM telemetry circuit. Neural Syst. Rehabil. Eng., IEEE Trans. **13**(3), 263–271 (Sept 2005)
93. K. Murari, C.M. Sauer, M. Stanacevic, G. Cauwenberghs, N. Thakor, Wireless Multichannel Integrated Potentiostat for Distributed Neurotransmitter Sensing Engineering in Medicine and Biology 27th Annual Conference Shanghai, China, Sept 1–4 2005
94. R. Thewes, F. Hofmann, A. Frey, B. Holzapfl, M. Schienle, C. Paulus, P. Schindler, G. Eckstein, C. Kassel, M. Stanzel, R. Hintsche, E. Nebling, J. Albers, J. Hassman, J. Schülein, W. Goemann, W. Gumbrecht, Sensor arrays for fully-electronic DNA detection on CMOS IEEE international solid-state circuits conference, Section 21.2, 2002
95. E. Ghafar-Zadeh, M. Sawan, D. Therriault, CMOS based capacitive sensor laboratory-on-chip: a multidisciplinary approach. Analog Integrated Circuits and Signal Processing. **59**(1), 1–12 (April 2009)
96. C. G. Zoski, *Handbook of Electrochemistery*, Elseiber, 2007, ISBN: 0-444-51958-0
97. C. Myung-suk, L. Sang-won, K. Jong-chul, C. Jun-dong, K. Jin-kwon; S. Hang-sik; L. Myung-ho; C. Un-sun; K. Jae-seok. Implantable Bio system design for displacement measurement of living life. Proc. 9th Int. Conf. Advanced Communication Technol. **1**, 12–14 299–304 (Feb 2007)
98. K. Kitamori, Micro and nano chemical system on chip. TRANSDUCERS 2007. International Solid-State Sensors, Actuators and Microsystems Conference, 2007. 10–14 June 2007 11–16
99. W. Chung, A.C. Paglinawan, Y. Wang, T. Kuo, A 600μW Readout circuit with potentiostat for amperometric chemical sensors and glucose meter applications. IEEE conference on electron devices and solid-state circuits, 2007. EDSSC 2007. 20–22 Dec 2007 pp. 1087–1090
100. A. Lasia. *Electrochemical impedance spectroscopy and its applications*, Modern Aspects of Electrochemistry. Vol. 32 (Kluwer Academic/Plenum Publishers, New York, 1999), Chapter.2, p. 143
101. R.J. Reay, S.P. Kounaves, G. Kovacs, An integrated CMOS potentiostat for miniaturized electroanalytical instrumentation, Digest of Technical Papers of the 1994 IEEE International 41st ISSCC Solid-State Circuits Conference, 16–18 Feb 1994 162–163
102. T.D. Strong, S.M. Martin, R.F. Franklin, R.B. Brown, Integrated electrochemical neurosensors, Proceedings of the IEEE International Symposium on Circuits and Systems, 2006. ISCAS 2006. (21–24 May 2006) p. 4 pp. 4110–4113
103. Martin, S.M., Gebara, F.H., Larivee, B.J., Brown, R.B. A CMOS-integrated microinstrument for trace detection of heavy metals. IEEE J. Solid-State Circuits. **40**(12), 2777–2786 (Dec 2005)
104. European Commission, Groundwater at risk: Managing the water under us, DG Environment, (2008)
105. I.M. Petayev, Plasma oxygen during cardioplumonary bypass: a comparison of blood levels woth oxygen present in plasma lipid. Clin. Sci. **94**(1), 35–41 (1998)
106. S.M. Park, J.S. Yoo. Electrochemical impedance spectroscopy for better electrochemical measurements. Anal. Chem. **75**(21), 455A–461A (2003)
107. A.D'. Amico et al., Low-voltage low-power integrated analog lock-in amplifier for gas sensor applications. Sens. Actuators B: Chem. (2009), doi:10.1016/j.snb.2009.01.046

108. M. Min, T. Parve, Improvement of lock-in electrical bio-impedance analyzer for implantable medical devices. IEEE Trans. Instrum. Meas. **56**(3), 968–974 (June 2007)
109. H.A. Wolpert, Use of continous glucose monitoring in the detection and prevention of hypoglycemia, J. Diabetes Sci. Technol. **1**(1), 146–150 (Jan 2007)
110. http://www.inhabitat.com/2008/12/11/tokyo-subway-stations-get-piezoelectric-floors/
111. http://www.sciencedirect.com/science?_ob=ArticleURL&_udi=B6TFC-4DFT3VG-3&_user=145085&_rdoc=1&_fmt=&_orig=search&_sort=d&_docanchor=&view=c&_searchStrId=1175746523&_rerunOrigin=google&_acct=C000012098&_version=1&_urlVersion=0&_userid=145085&md5=5faacc341698b061b7db5d4fc1215f28 – bbib3
112. L.A. Cantarero, J.E. Butler, J.W. Osborne, The adsorptive characteristics of proteins for polystyrene and their significance in solid-phase immunoassays Anal. Biochem. **105**, 375–382 (1980)
113. M.M. Teymoori, H. Asadollahi, MEMS Based Medical Microsensors Computer and Electrical Engineering, 2009. ICCEE '09. 2nd International Conference on Vol. 1 (28–30 Dec 2009) pp. 158–162
114. Y.J. Lee, J.D. Kim, J.Y. Park, Flexible enzyme free glucose micro-sensor for continuous monitoring applications. Solid-State Sensors, Actuators and Microsystems Conference, 2009. TRANSDUCERS 2009. International 21–25 June 2009 pp. 1806–1809
115. J.D. Goud, P.M. Raj, Jin Liu, R. Narayan, M. Iyer, R. Tummala, Electrochemical Biosensors and Microfluidics in Organic System-on-Package Technology Electronic Components and Technology Conference, 2007. ECTC '07. Proceedings 57th. May 29–June 1 2007 pp. 1550–1555
116. E. Carnes, E. Wilkins, The development of a new, rapid, amperometric immunosensor for the detection of low concentrations of bacteria part I: Design and detection system and applications. Am. J. Appl. Sci. 2 and 3, 597–606 (2005)
117. H. Lhermet. et al., Efficient Power Management Circuit: Thermal Energy Harvesting to Above-IC Microbattery Energy Storage, ISSCC Digest of Technical Papers, pp. 62–63, Feb 2007
118. H. Shao et al., An inductor-less micro solar power management system design for energy harvesting applications. IEEE international symposium on circuits and systems, May 2007, pp. 1353–1356

Chapter 2
Energy Harvesting (Multi Harvesting Power Chip)

Abstract There is a growing interest in renewable energy and their applications for both, high and low power systems. Specifically, Energy Harvesting consists in the use of free available energy from the environment, vibrations, heat, light, radio waves or human activities, to power small electronic systems with Low-Voltages and Low-Power consumption. The challenge is to avoid the use of any bulky battery with finite amount of energy and just work directly with the harvested energy and a rechargeable storage element. At that point, Energy Harvesting generators are a promising alternative to generate energy from environment sources and power some applications. Moreover, the use of these generators, with infinite amount of energy, allows the development of autonomous Self-Powered applications. This chapter discusses the development of a real power system based on the recollected energy from several ambient sources. A system able to collect and manage energy from four different power sources, solar light, vibrations, thermal and inductive waves is introduced. Furthermore, the conception is validated with a full-custom Integrated Circuit (IC). Later on, a comprehensive description of all circuits involved in the Multi harvesting system is done; emphasizing the design for Low-Voltage and Low-Power applications.

Keywords Electric power generation · Energy scavenging · Low-power electronics · Power conditioning

2.1 Multi Harvesting Power Chip (MHPC)

MHPC consists on a simple Integrated Circuit (IC) able to harvest energy from different ambient sources and deliver this energy into usable electrical form [1, 2]. The proposed architecture is presented in Fig. 2.1.

The system is able to collect energy from three different ambient sources: Solar (SC), Mechanical Vibration (PZT) and Thermoelectrical energy (HEAT). Furthermore, it is also possible to recollect energy from an Inductive power link (MI). All the energy obtained from these four sources is then transformed into electrical energy and stored in one or more storage devices (SD).

J. Colomer-Farrarons, P.L. Miribel-Català, *A CMOS Self-Powered Front-End Architecture for Subcutaneous Event-Detector Devices*, DOI 10.1007/978-94-007-0686-6_2,

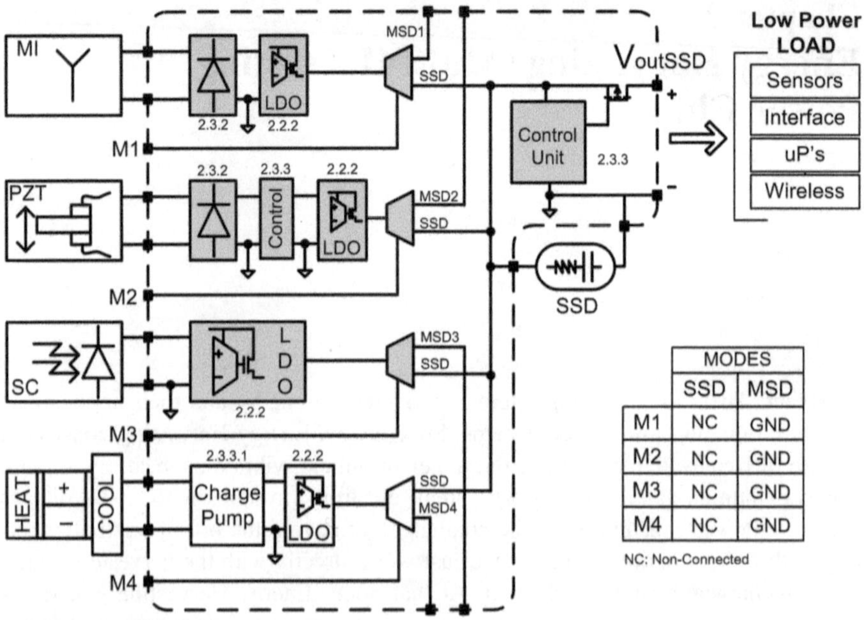

Fig. 2.1 Multi Harvesting Power Chip (MHPC) architecture

The objective is to combine all the energy sources at the same time [3–12]. In that way, all individual energies are added with the other ones obtaining a summation of energies. Moreover, this multi-power source scheme is very robust against power failures due to the nature of their conception; MHPC uses several independent power sources in one IC and, when one or more doesn't work properly, the system still produces energy thanks to the other sources. Each power source uses its own energy conditioning circuit, in that way, all the sources can work together, in parallel and at the same time.

There are two operation modes when working with MHPC. The first one consists on using just a Single Storage Device (SSD) shared with all sources. The second mode consists in the use of each source regardless of the other ones. In that mode, it is necessary to use Multiple Storage Devices (MSD), one for each source. Using the second mode it is possible to power more than one independent load at the same time, whereas using the first mode, it is possible to collect more energy in the storage device but only power one common load.

Another important aspect consists in controlling the stored energy and how to transfer it to the load. Basically, the MHPC introduces a control module that manages this situation. In an SSD operation mode, the control senses continuously the voltage at the storage element and, when there is enough voltage, the load is connected. At that point, if the voltage decreases till a minimum value, also controlled by the control module, the load is disconnected. This methodology is explained in detail in Section 2.3.3.2. On the other side, in MSD mode, it is possible to integrate one individual control module for each source or just connect directly the source with the load.

Regarding the control module and the conditioning circuits, it is logic to assume that a low power consumption design is mandatory. Since a Multi Harvesting Power Chip is a Self-Powered device and focused on low-voltage applications, a low consumption is mandatory for all circuits in order to obtain good efficiency values. This statement implies that non-complex power management circuits with low power architectures and circuits have to be specifically designed.

A full custom IC has been designed to validate the MHPC conception. The designed system is able to work with three different sources: Solar Energy (SC), Mechanical Vibration (PZT) [13] and Magnetic Induction Link (MI). The implemented integrated modules are colored in gray in Fig. 2.1 (Thermoelectric Source is not integrated due to area limitations; the integration of an autonomous efficient DC/DC step-up converter, which is needed for such type of sources, like a Charge-Pump implies high area space. Anyway, the validation can be done without it.)

The device implements both operation modes: SSD and MSD. Signals M1–M4 control the connection of each power source. A control module is included in order to work in SSD mode. It is based on the Simple Control Module architecture presented in Section 0 and modified to work between 1.2 V and 1 V (1.2 V is the LDO regulated voltage). The adopted control option in Multiple Storage Device (MSD) mode is just to connect directly the load with the source. Figure 2.2 depicts the final IC layout.

The IC has been tested with the Demo-Board presented in Fig. 2.3. Six solar cells XOB17 from IXYS [14] are used to harvest solar energy. Each one of them could be connected with the others using a serial or parallel connection. Each solar cell produces 1.5 V and 11 mA in the P_{mpp} point. An electrical model of the solar cell is presented in Section 2.2.1.

Several Piezoelectric transducers are used to handle with the mechanical vibrations. The main tests were done with two piezos from Mide. Tech®: QP20W and QP40W [15]. The models and main features of all those piezoelectrics are obtained from [16, 17] and the original datasheets. Section 2.3.1 introduces the models used in the simulation. A detailed explanation of piezoelectric modeling is out of the scope of the present work. More accurate information about piezoelectric transducers modeling can be found in [10, 18, 19].

Regarding Magnetic Induction link, a planar rectangular antenna (Inductor) of 30 mm × 15 mm is added to the board in order to work as a secondary antenna. It has three turns, a conductor width of 1 mm and an inductance of 220 nH. It is tuned to work in the RFID range of 13.56 MHz. A Texas Instruments TRF7960 [20] is used as primary source. A model of the Inductive Link as well as the secondary inductor can be developed following the procedures introduced in [21, 22]. A simple introduction to inductive link modeling is done in Section 2.2.1. For more information just refer to references.

Table 2.1 shows the main features of the MHPC in terms of its electrical characteristics. These are the minimum and maximum voltages and currents available

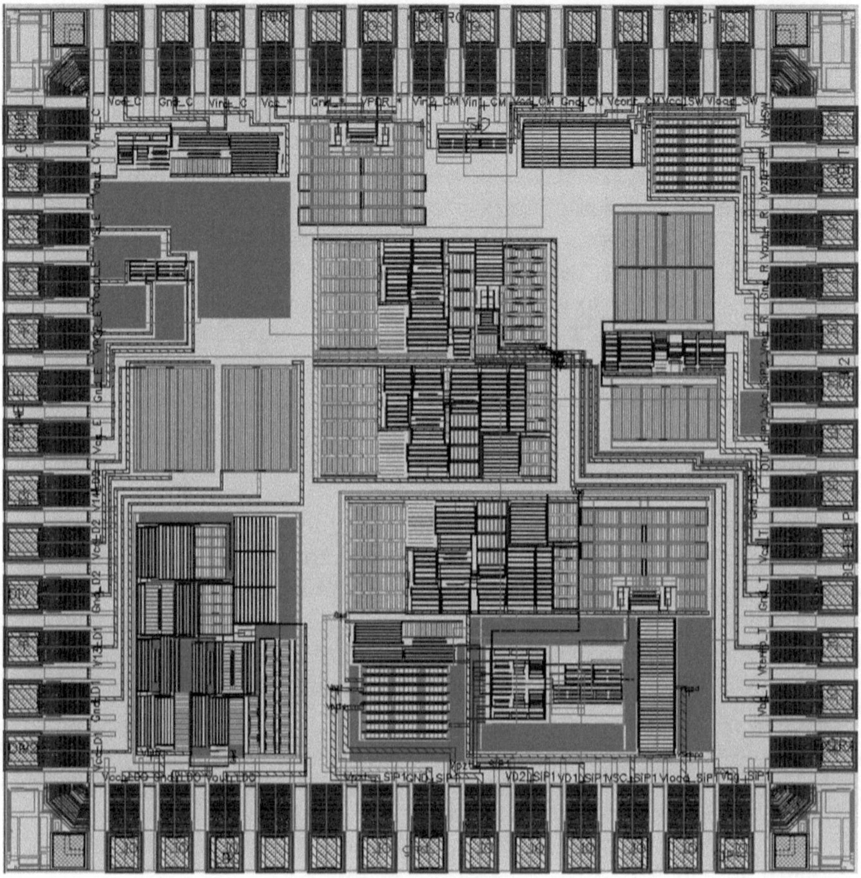

Fig. 2.2 Multi Harvesting Power Chip (MHPC) IC layout

for this prototype for the three considered types of powering source: vibration based on piezoelectric generators (Pzt), light based on indoor solar cells (SC) and, finally, electromagnetic induction (RF). On the other hand, Table 2.2 presents the electrical features of the prototype for the MSD mode, which is that every powering source has its own storage device, or the SSD mode, where the three powering channels share the same storage device, which can be a super capacitor. In our case, as a test approach for Fig. 2.18, a capacitor of 47 μF was used.

Finally, Fig. 2.4 illustrates the working principle of the MHPC operation in SSD mode, one Storage Device shared with all sources. The current used to charge the Storage Device (SD) is depicted and how it increases or decreases depending on the number of activated sources.

The voltage at the SD is regulated at 1.2 V. The maximum current flowing to the SD is 5 mA when the three sources, Solar (SC), Piezoelectric (PZT) and Inductive Link (MI), are working together. This particular test is done with the

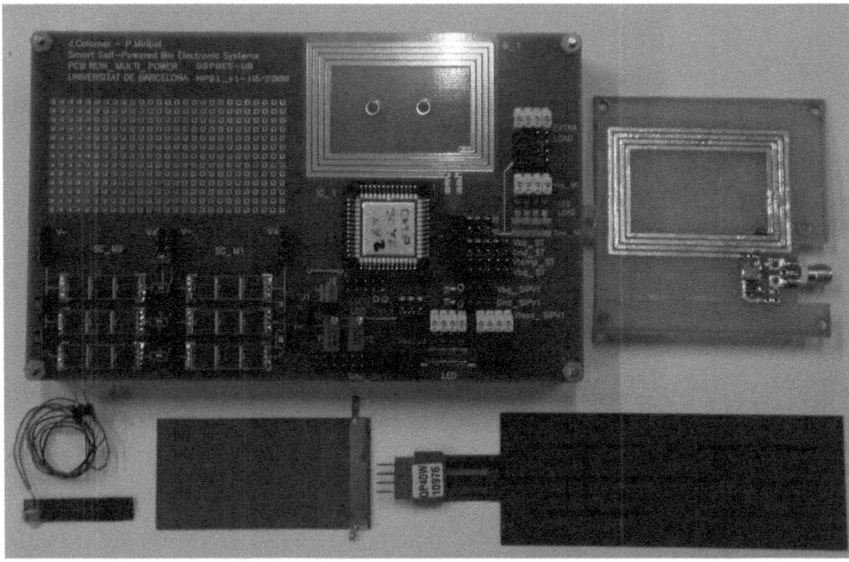

Fig. 2.3 MHPC demo board

Table 2.1 MHPC Input pins maximum electrical features

Symbol	Parameter	Min	Max	Units
$V_{IN}PZT$	Input voltage piezo	± 1	± 2.5	Vac
I_{IN} PZT	Input current piezo	–	5	mA
V_{IN} SC	Input voltage solar cell	1.3	2.5	Vdc
I_{IN} SC	Input current solar cell	–	10	mA
V_{IN} MI	Input voltage induction	± 1.3	± 2.5	Vac
I_{IN} MI	Input current induction operation freq. up to 25 MHz		10	mA

Table 2.2 MHPC Output pins maximum electrical features in function of the operation mode

	Symbol	Parameter	Min	Max	Units
MSD mode	V_{OUT} PZT	Output voltage	1	2.5	Vdc
	I_{OUT} PZT	Output voltage	–	10	mA
	V_{OUT} SC	Output voltage	–	1.2	Vdc
	I_{OUT} SC	Output voltage	–	10	mA
	V_{OUT} MI	Output voltage	–	1.2	Vdc
	I_{OUT} MI	Output voltage	–	10	mA
SSD mode	V_{OUT}	Output voltage	1	1.2	V
	I_{OUT}	Output voltage	–	20	mA

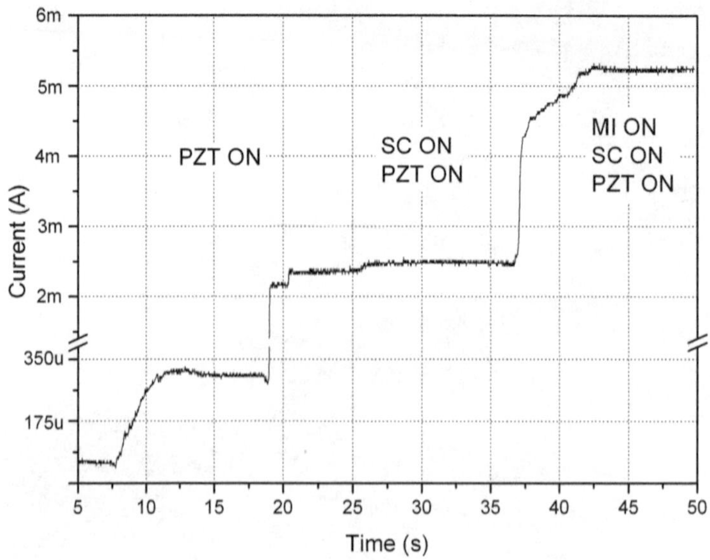

Fig. 2.4 MHPC storage device current evolution in function of the energy sources

QP40W piezoelectric working at 1 m/s2 – 80 Hz, 2 IXYS solar cell at indoor light condition of 1,500 lx and the RF link generator TRF7960 emitting at full-power (200 mW).

The following sections explain in detail the circuits used in the development of the integrated Multi Harvesting Power Chip. Figure 2.1 also indicates the section number where each module is explained.

2.2 Solar and Inductive Power Harvesting

This section explores the management of the energy provided by the wireless power link and by the energy generated by solar radiation [23]. As it is well known, the solar energy can be collected using photovoltaic cells or panels [24]. Those cells generate an unregulated DC voltage when light falls on them so, a DC regulator is necessary to stabilize the voltage. On the other side, an inductive power link generates an unregulated AC voltage at the secondary side of the link. This means that a rectifier is firstly necessary to generate an unregulated DC voltage and then, the DC voltage has to be regulated.

It is also possible to introduce an extra stage in the solar cell power scheme to increase the efficiency of the solar generator. This stage tries to work continuously in the M_{PP} (Maximal Power Point) of the solar cell. That module is not implemented in the present work because it is well solved in the literature; there are several good low-power approaches to implement this kind of stages like in [25].

Both power schemes are depicted in Fig. 2.1 and, as can be observed, they use the same regulator (LDO). The adopted regulator is based on the design of a Low-Dropout linear regulator [26, 27] because it presents a good trade-off in terms of conversion efficiency and power consumption. A discussion of the LDO conception and design is done in the following of this chapter. Moreover, the design of a precise Bandgap voltage reference circuit is also presented. This circuit has an important role on the LDO operation and, even, in the circuits discussed further on.

2.2.1 Solar and Inductive Models

2.2.1.1 Solar Cell Model

A solar cell can be easily described as a large area semiconductor diode (P-N junction) able to convert into electrical current the energy of the incident light (energy of photons) thanks to the photovoltaic effect [28].

The basic equivalent circuit model of a solar cell is formed by a diode (D_1) and a current source (I_1) connected in parallel. The current source is proportional to the light radiation whereas the diode represents the P-N junction.

Moreover, two resistors, one in series (R_S) and other in parallel (R_P), could also be introduced in order to take into account the imperfections of the solar cell. R_S resistance is needed because a solar cell is not a perfect conductor and R_P models its internal leakage current [29]. In an ideal cell, it is logic to assume $R_S = 0$ and $R_P = \infty$. Sometimes an extra capacitor (C_1) is added to simulate the parasitic capacitance of the junction. Figure 2.5 presents a solar cell model.

A solar cell can also be described with its I-V waveform. Figure 2.6 presents the I-V characteristics for a solar BIT from IXYS [14]. The generated current and voltage varies in function of the light conditions although it is important to describe three quantities:

- Open circuit voltage (V_{oc}): The voltage between the terminals when no current is flowing (infinite load resistance).
- Short circuit current (I_{sc}): The current when there are a short-circuit in the terminals (zero load resistance).
- Maximal Power Point (M_{PP}): Point of the solar cell when the maximum amount of power is produced.

The IXYS XOB17 solar cell is used in that work as a solar generator. It is a high efficiency indoor / outdoor Monocrystal silicon solar cell mounted in a smd

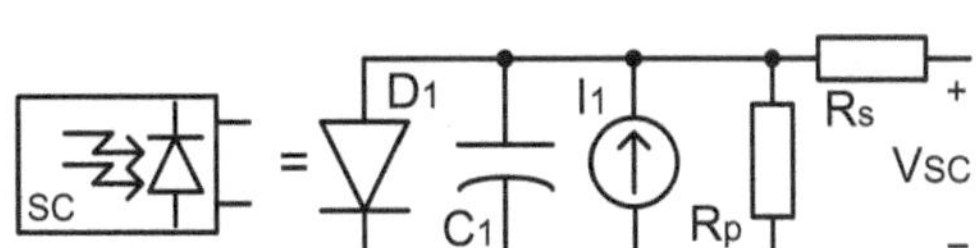

Fig. 2.5 Solar cell model and symbol

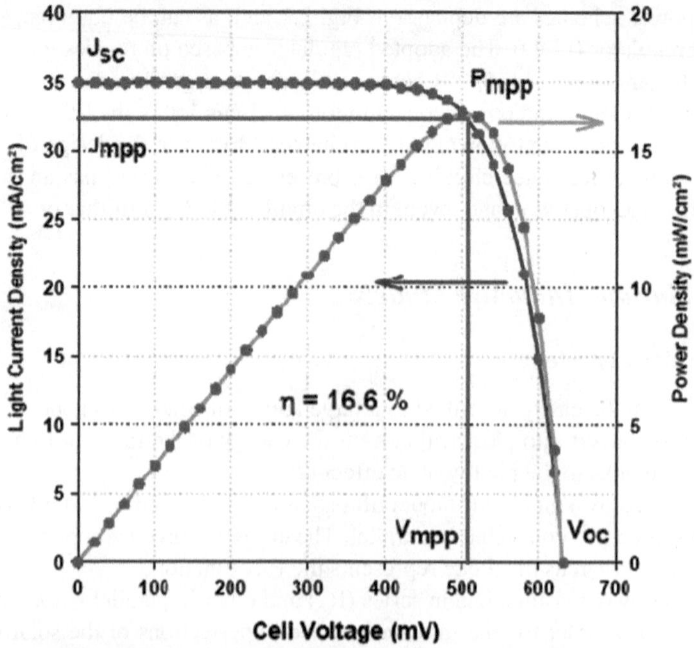

Fig. 2.6 I-V characteristic for a solarBIT from IXYS [14]

Table 2.3 IXYS XOB17 electrical characteristics (3 solar cells in series)

Symbol	Parameter	Value	Units
V_{OC}	Open circuit voltage	1.89	V
I_{SC}	Short circuit current	12.6	mA
V_{MPP}	Voltage @ MPP	1.53	V
I_{MPP}	Current @ MPP	11.7	mA

mechanical package (22 mm × 7 mm × 1.6 mm) formed by three XOD17 solar cells dies. It presents an efficiency of typically 17% and the characteristics of Table 2.3.

2.2.1.2 Inductive Power Link Model

Wireless Power Link, also known as Wireless Power Transmission, is the way to transfer energy, so it is the manner to transfer electricity through an electromagnetic field [30, 31, 32, 33]. There are several types of Wireless Power Transmission, radiative [34], conductive [35], capacitive [36] and inductive but, this work is only focused on Inductive Power Transmission.

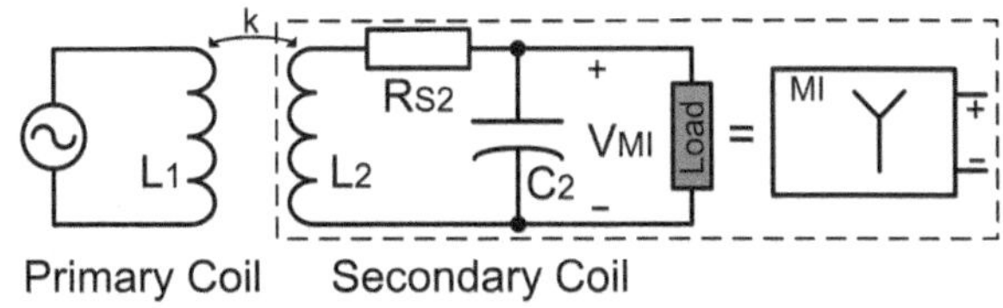

Fig. 2.7 Inductive power link model and secondary coil symbol

An Inductive Link is mainly composed by two coupled coils, Fig. 2.7, one works as a primary or transmitting coil (L_1), where the power is injected and emitted, and the other works as a secondary or receiving coil (L_2), where the total or a partial part of the emitted power is received.

The transmission principle consists in injecting an AC current through the primary coil generating an AC magnetic field. The secondary coil then recovers this magnetic field, or just a portion of it and an electromotive force (emf) is generated, which is used as a voltage source by the load connected to the receiving coil. This transmission method is also named transcutaneous energy transfer system [21] when the links is used in human implantable devices.

Moreover, inductive transmission presents some advantages if one of the possible applications is focused on the human body and implantable electronics. It is the best solution to develop a near field link [21] in the range of 1–15 MHz [36, 37] to transmit energy through the human skin. This range of frequencies assures minimal losses due to the power absorption of the skin and since the coils are coupled and tuned to work specifically at one frequency, the electrical drifts are minimal [22] to the surrounding biological area.

The secondary coil usually works in resonance to increase the link efficiency and to be selective with the used frequency. The resonance is achieved by adding a capacitor in series or parallel to the secondary coil. If the link uses a parallel-resonance secondary coil, it works as an output voltage source, whereas the series one operates as a current source.

The designed coils work in parallel resonance by adding an extra capacitor (C_2) as it is depicted in Fig. 2.7. Parallel resonance presets better performance for high loads than the series one. Furthermore, the parasitic capacitance of the secondary coil (L_2) is absorbed by the tuning capacitor. The resonant frequency of the L-C tank, no load connected, is defined by Eq. (2.1).

$$w_2^{Ideal} = \frac{1}{\sqrt{L_2 C_2}} \tag{2.1}$$

This equation is used to calculate the tuning capacitor for a specific given frequency. It is important to notice that the resonance frequency is modified when the load is connected to the secondary coil as it is described in Eq. (2.2). As a result, for low load values, high current, the resonant frequency is widely modified, hence low efficiencies. So, the parallel resonance secondary circuit is suitable to be used with large loads [21].

$$w_2^{Real} = \frac{1}{\sqrt{L_2 C_2}} - \frac{1}{\sqrt{R_{Load}^2\, C_2^2}} \tag{2.2}$$

The coupling factor k between coils is another important parameter to take into account. The power link efficiency is mainly dominated by k. Horizontal and vertical misalignments as well as large distances between both coils affect the coupling factor reducing the efficiency of the link. The total link efficiency expression for secondary parallel resonance is given by

$$\eta = \frac{k^2 Q_1 Q_2}{\left(1 + \dfrac{Q_1}{wC_2 R_{Load}} + k^2 Q_1 Q_2\right)(wC_2 R_{Load} + Q_1)}$$
$$Q_1 = \frac{wL_1}{R_{S1}}\ ;\ Q_2 = \frac{wL_2}{R_{S2}} \tag{2.3}$$

where Q_1 and Q_2 are the quality factor of the primary and secondary coils respectively. The efficiency is independent regardless if the primary coil is tuned or not. The maximal efficiency is obtained for

$$R_{Load}^{\eta MAX} = \frac{Q_2}{\left(1 + \sqrt{1 + k^2 Q_1 Q_2}\right) wC_2} \tag{2.4}$$

Then, the maximal link efficiency is just substituting Eq. (2.4) in Eq. (2.3):

$$\eta_{MAX} = \frac{k^2 Q_1 Q_2}{\left(1 + \sqrt{1 + k^2 Q_1 Q_2}\right)^2} \tag{2.5}$$

A rectangular coil has been designed in a Printed Circuit Board (PCB) in order to work as a secondary coil for harvesting applications (Fig. 2.3). It is tuned at the RFID frequency of 13.56 MHz. A Texas Instruments TRF7960 is used as primary source. Table 2.4 summarizes its main characteristics.

An accurate description and a calculation methodology to develop an inductive link for harvesting and implantable devices can be found in references [21, 22]. In the same way, references [38, 39, 40] explain several options to design different types of coils.

Table 2.4 Rectangular coil characteristics

Symbol	Parameter	Value	Units
L_2	Secondary coil	220	nH
C_2	Parallel capacitor	620	pF
R_{S2}	Secondary series resistance	310	mΩ
f	Resonance frequency	13.56	MHz
N	Number of turns	3	-
CW	Conductor width	1	mm
SUB	Type of substrate	FR4	-
L x W	Dimensions	30 × 15	mm

2.2.2 LDO Regulator and Bandgap Reference Circuit

The implemented Low-DropOut (LDO) regulator architecture is presented in Fig. 2.8. It is formed by one Switch, one Error Amplifier (EA), a voltage Bandgap (BG) reference circuit and the close-loop resistors (CLR) used to regulate the output voltage. It is designed to generate a nominal regulated voltage (V_{Reg}) of 1.2 V. Generally, the regulated voltage is defined by the CLR as:

$$V_{Reg} = V_{BG}\left(1 + \frac{R_1}{R_2}\right) \tag{2.6}$$

The selected switch is a PMOS transistor [41]. An NMOS switch implies positive voltages to drive its gate and the signals generation becomes difficult when working at low voltages. The low voltage control and low dropout voltages of the PMOS transistor are perfect for this application.

The design of the LDO has to follow these two main constrains: the lowest consumption and smallest area as possible. The consumption depends on the power dissipation of the internal circuits like the current source, the voltage reference circuit and error amplifier. In terms of area, it is logic to assume that the smallest area as possible is the desirable solution to integrate the LDO in a system like MHPC.

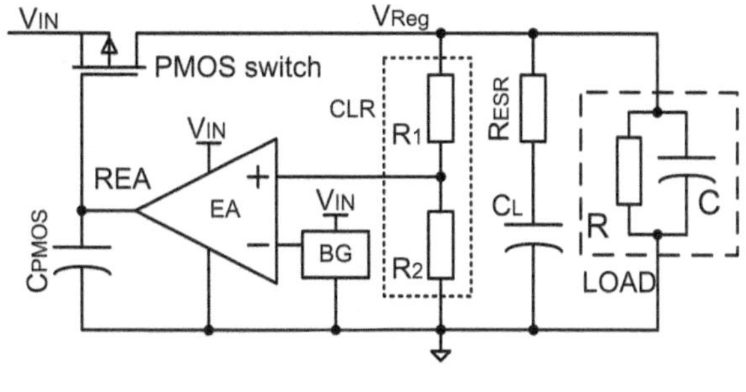

Fig. 2.8 LDO regulator architecture

It is also important to take care with the DC load currents. They can vary from the ground current, defined by the minimum current used by the LDO circuits, around 20 μ A, and the maximum load current condition defined as 10 mA. Furthermore, the load current variation plays an important role in the stability. The LDO has to be stable in all the DC load current range.

Taking into account the stability issues and the power and area constrains the LDO has been designed following the idea of a compensation by an external dominant pole [42, 43, 44]. This means the LDO will use an off-chip capacitor (C_L) to guaranty the stability for all load conditions.

To analyze the stability it is important to evaluate the position of the poles and zeros of the LDO. The generic LDO has three poles in its frequency loop transfer. The first pole (P_1) is basically defined by the output resistance of the switch, the load resistance and, the series resistance (R_{ESR}) of the output capacitor C_{LOAD}, Fig. 2.8.

The second pole (P_2) is related with the gate capacitance of the PMOS switch and the output resistance of the Error Amplifier (REA). The first pole, P_1, is the dominant one. This pole has the characteristic that varies depending on the load current level in low frequencies range. Finally, the ESR resistance and the equivalent load capacitance define the third pole, P_3.

Typically the REA resistance is high and the $C_{PMO}S$ is small; hence P_2 is also placed at medium/high frequencies, which constrains the response of the LDO to be slow. But in this system there is also the presence of a zero, which is typically related with the R_{ESR} resistance. In that way, the zero is located at higher frequencies than P_1 and, at the same time, at lower frequencies that the Unity Gain Frequency (UGF) pushing the LDO in a stable situation.

So, in summary, there is a dominant pole, P1, working at low frequencies and another pole P_3 working at higher frequencies than UGF. Furthermore, there is P_2 working at medium/high frequencies but lower than UGF, which makes the stability difficult. For that reason, the placement of the zero (Z_0) at lower frequencies than UGF helps the stability of the LDO.

The stability of the LDO by a dominant pole is achievable but large value of C_L is necessary. To obtain a high value capacitor is not difficult when working with harvesting applications because the storage element usually presents a high capacitance value. The speed of the LDO is typically of several ms (around 25 ms).

The Error Amplifier is based on a two-stage amplifier [44] as it is depicted in Fig. 2.9. Transistors M_{N1} and M_{N2} are used to define the ground current in the case of no-load conditions.

The LDO is prepared to work with an input range from 1.3 V up to 2.5 V. However, it presents its better performance in the range of 1.3–1.8 V. Figure 2.10 (left) presents the line regulation for a V_{IN} variation from 1.4 V to 1.6 V and a fixed load of 10 kΩ / 10 μF. Otherwise, Fig. 2.10 (right) depicts the load regulation for a fixed voltage of 1.4 V and a load variation from 1 MΩ to 1 kΩ. Table 2.5 summarizes the most important features of the implemented LDO [45].

Figure 2.11 shows the experimental case of a regulated voltage (V_{Reg}) of 1.196 V obtained from an Unregulated DC voltage of 1.5 V provided by the solar cells at an indoor illumination conditions of 1,500 lx (Fluorescent Tubes). A 4.7 nF external

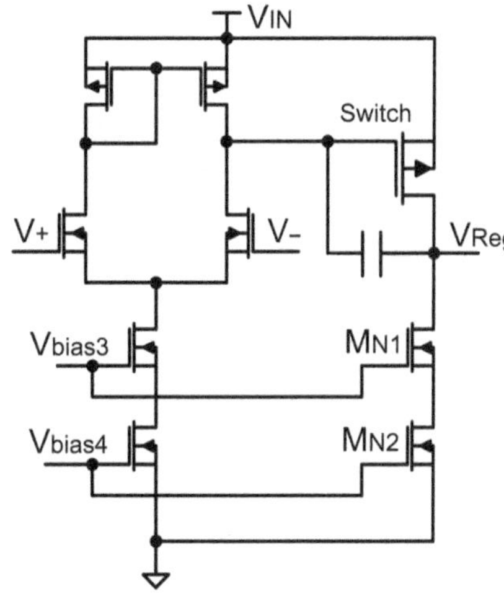

Fig. 2.9 Error amplifier and PMOS switch

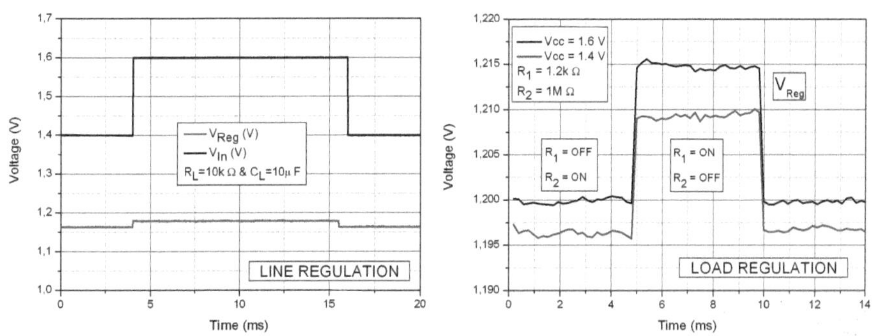

Fig. 2.10 LDO Line (*left*) and load (*right*) regulation

Table 2.5 LDO electrical characteristics

Symbol	Parameter	Min	Max	Units
V_{IN}	Input voltage	1.3	2.5	V
V_{Reg}	Regulated output voltage	1.189	1.22	V
I_{Reg}	Output current	20 μ	10 m	A
I_{CC}	Current consumption	23 μ	27 μ	A
P_{CC}	Power consumption	29 μ	67 μ	W
Λ_{Load}	Load regulation*	13 m	34 m	V
Λ_{Line}	Line regulation**	7 m	18 m	V
Λ_{DC}	DC gain	63	72	dB
P_M	Phase margin	58	65	°
PSSR	Power supply rejection ratio	28.7	39.4	dB

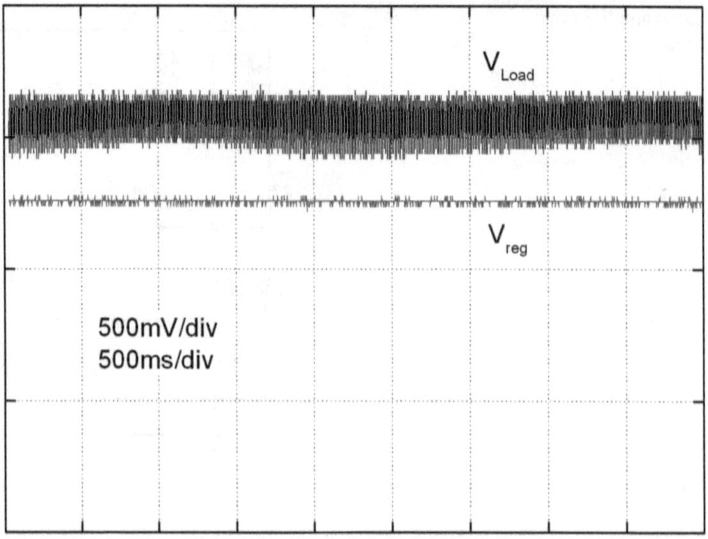

Fig. 2.11 Solar cell + LDO experimental result

capacitor (C_L) works as storage device and to compensates the LDO. Further on, the LDO is used to regulate the voltage obtained with an inductive link working at 13.56 MHz.

2.2.2.1 Bandgap

The proposed Bandgap reference circuit is based on a full MOSFET Bandgap architecture [46, 47], in order to avoid the use of bipolar transistors [48, 49]. There are not available BJT transistors in the used tech. The full MOSFET Bandgap architecture is depicted in Fig. 2.12. This topology is really interesting because it presents a good trade-off between performance, area and power consumption.

This Bandgap consists in a self-biased peaking current source [50, 51, 52] with a series resistor. Transistors M_3, M_4 and resistor R_1 form the peaking reference

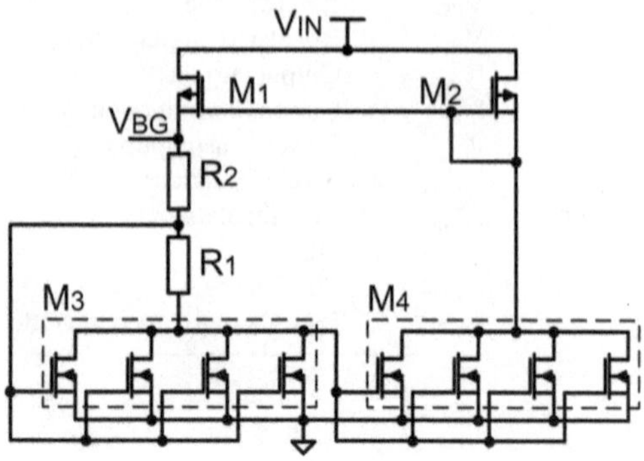

Fig. 2.12 Schematic of a BG circuit

current source. The transistors work in the sub-threshold region. The transistor M_4 is designed such that I_{D4} is at its peaking value. The design condition is defined by

$$I_{D3}R_1 = NV_T \tag{2.7}$$

where N is the ideal factor and V_T is the thermal voltage (kT/q). Then, can be demonstrated that the relationship between I_{D3} and I_{D4}, if the last condition is satisfied, is:

$$I_{D4} = I_{D3}\frac{K_4}{K_3}e^{-1} \tag{2.8}$$

where $K_3 = W_3/L_3$, and $K_4 = W_4/L_4$. Towing to the current mirror, the condition $I_{D3} = I_{D4}$ is defined, then the relationship between K_3 and K_4 to satisfy the last condition is $K_4/K_3 = e$. If this ration is defined between the transistors the peaking condition is maintained.

The currents are proportional to the absolute temperature (PTAT). Then, the voltage drop in R_1 is PTAT. The Bandgap reference voltage (V_{BG}) is defined by

$$V_{BG} = \frac{R_2}{R_1}NV_T + V_{GS3} \tag{2.9}$$

with the correct ratio between R_2 and R_1 it is possible to compensate the variations due to the temperature of V_{GS3}.

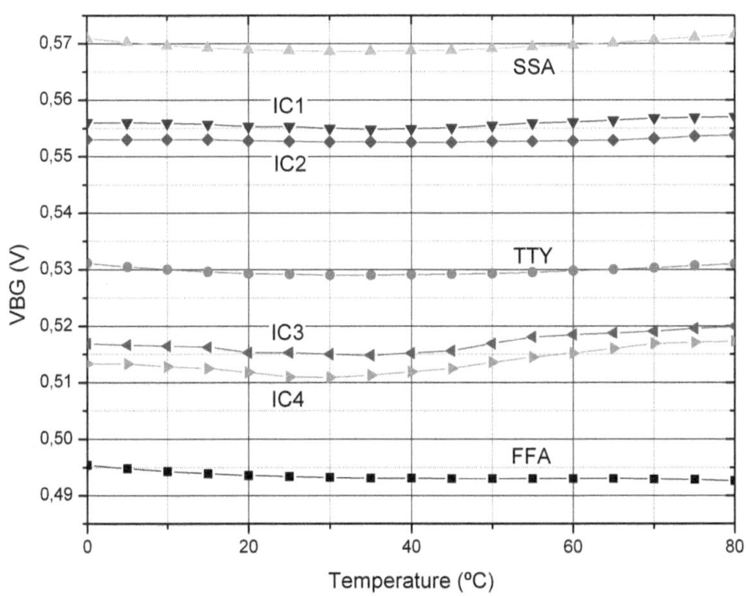

Fig. 2.13 Bandgap voltage vs. temperature for four different IC

Several simulations have been carried out obtaining a nominal value of the reference voltage of 530 mV. The temperature coefficient is 100 ppm / °C, and a PSRR of 42 dB can be expected for T = 27 °C and $V_{CC} = 1.3$ V. The PSRR is 80 dB for $V_{CC} = 1.8$ V, and even greater (~90 dB), for higher values up to 2.5 V. The power dissipation is 11 μW @ 1.4 V, which represents low power consumption.

Figure 2.13 presents the experimental Bandgap voltage vs. temperature at $V_{CC} = 1.2$ V. The results show the experimental values of four different circuits, I_{C1}, I_{C2}, I_{C2} and I_{C4}. Also, the experimental voltages are compared with the simulated BG voltages for different statistical models, SSA (Slow case), TTY (Typical case) and the FFA (Fast case). The BG starts to work properly at 800 mV. Some statistical Montecarlo analyses are presented in Appendix A.

2.3 Piezoelectric Harvesting

As it is well known, piezoelectric membranes generate an unregulated AC voltage. Thus, they cannot be connected directly with the load because it generally requires a DC voltage. Therefore, three parts usually form a power scheme based on piezoelectric generators, Fig. 2.14.

The first stage is the piezoelectric generator or micro generator –MPG- (Section 2.3.1) and its AC voltage, which is used as the main power source. The second stage is the Rectification (Section 2.3.2) that converts the incoming AC signal wave of the PZT into a non-regulated DC voltage. This stage is an important step because it has to extract the maximum voltage and power from the piezoelectric and deliver them without considerable losses to the next stage. Finally, the Regulation stage (Section 2.3.3) regulates and controls the non-regulated DC voltage coming from the second stage in order to transfer the energy to the load.

The basic working principle is simple; an AC voltage is generated when a mechanical vibration is applied on the piezoelectric. This voltage is automatically rectified by the second stage and converted into a non-regulated DC voltage. Meanwhile, the capacitor (CR) is charged. Later on, the regulation stage handles the voltage at the capacitor and delivers it to the load.

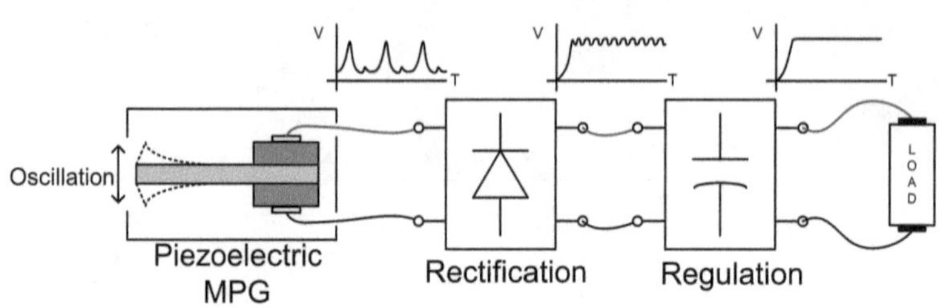

Fig. 2.14 Piezoelectric harvesting general architecture

There are several solutions to design each stage, from the simple diode bridge rectifier for the Rectification stage to the classical DC/DC converters for the third stage. The problem is that many of these solutions are focused on high voltage generators [53, 54] and they could not be applied for Low-Voltage Low-Power systems due to losses involved in each stage. For instance, the drop voltage in a diode can be neglected in those applications where the generator produces a voltage higher than 10 V. However, if the generated voltage is in the range of 1–2 V the diode forward voltage represents an important voltage percentage of the incoming signal, so, this situation implies low efficiencies.

This section introduces several rectifiers' topologies as well as two regulation methods used in Low-Voltage applications focused on the development of Self-Powered Systems. It is desired to avoid the use of an external battery and just power the whole system with the energy provided by the generator. Furthermore, a fast introduction to piezoelectric generators is also introduced.

2.3.1 *Piezoelectric Model*

The main Piezoelectric generator used is based on the Quick Pack QP20W (Midé Technology Corporation, Medford, MA, USA). The QP20W is a composite beam made of 2 piezoelectric layers working as a bimorph, with an intermediate layer based on Polyimide. This composite beam is then located with one end clamped to a vibrating body and the other end remaining free. The vibrations forced at the clamped end are propagated along the cantilever beam. This wave generates an induced strain in the membrane, which at the same time produces an electrical charge.

In order to be able to simulate this energy recovery, a Lumped Electrical Model compatible with Spectre software has been developed, Fig. 2.15. The model is based on the Modal analysis of piezoelectric Euler-Bernoulli beam equations [55, 56, 57]. Solving the beam equation for the first resonance mode, and considering the equivalences between the equations defining a mechanical system and the equations defining an electrical circuit, an electromechanical equivalent circuit is proposed. Each of the electrical components, appearing in the equivalent circuit, has been defined by means of the beam geometrical and material parameters. Due to the composite nature of the material some of these parameters are experimentally identified [58, 69, 57].

In the experimental setup for model identification and validation the transducer is mounted and clamped over an electromagnetic vibration shaker ET-132 Labworks®

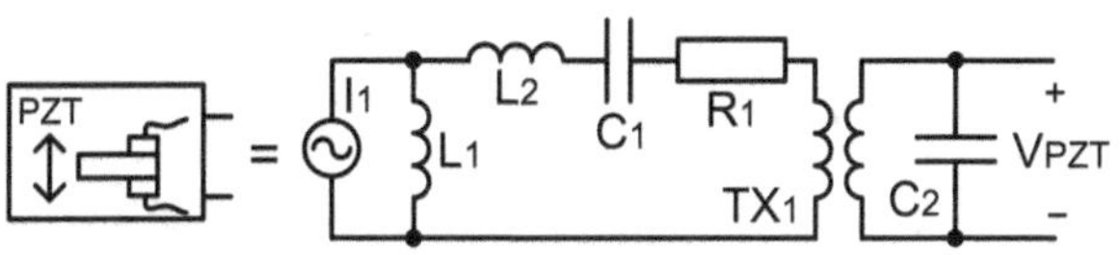

Fig. 2.15 PZT electrical model

Table 2.6 QP20W electrical characteristics

Acceleration (m/s^2)	Voltage (V)	Current (A)	R_{LOAD} (Ω)
7 @ 82 Hz	1.2	75 μ	10 k

Inc. The shaker is driven with a specific amplifier Pa-119 Labworks® Inc., with the command signals coming from a function generator. For the displacement measurements, a triangulation laser LC2440 from Keyence® is used.

This model of the PZT is presented in Fig. 2.15 [60]. Table 2.6 summarizes the features for this model, just in the case of an output voltage of 1.2 V. The maximum voltage available is 2.5 V, requested by the technological limitations. In the same way a model for the Quick Pack QP40W is used [59] in order to increase the current capability of the power source, different PZT have been used at the same time. This new power source structure uses in parallel different PZT allowing the system to obtain a higher current.

Different simple PZT generators are connected in parallel creating an array of generators. This array is then connected to the power conditioning circuitry to charge the Storage Device. This topology allows the use of one or several PZT generators increasing or decreasing the total current generated by the PZT's.

Furthermore, the use of several PZT transducers in parallel decreases the time required to charge the Storage Device in the initial start-up. This time reduction can be observed in Fig. 2.16 where different charging times for different

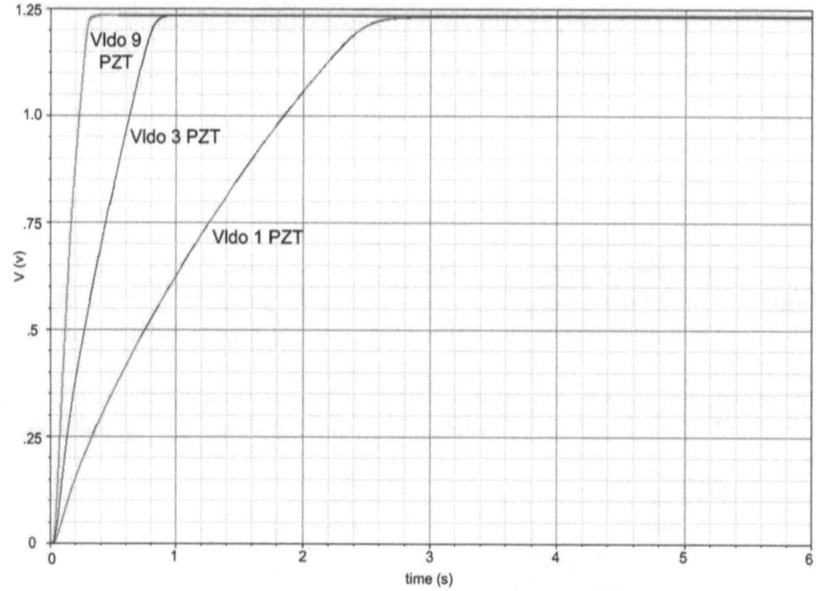

Fig. 2.16 Simulated charge time for different arrays of QP20W. It is assumed that all the PZT are vibrating simultaneously at the same frequency

Table 2.7 PZT array current generation

PZT in parallel	Current (A)	Charging time CL = 100 μF (s)
1	75 μ	2.6
3	120 μ	1
9	300 μ	0.380

arrays of QP20W generators using a 100 μF capacitor are depicted. Waveform Vldo9 represents the charging time for an array composed by 9 simple GP20W generators.

Waveform Vldo3 represents the time involved in the charge for an array of 3 elements, whereas waveform Vldo1 represents the time for an array of 1 PZT. This simulation has been carried out to show how the use of an array of simple PZT can reduce the charge time. Table 2.7 summarizes the features for the three used different arrays of PZT (It is assumed that all PZT are working simultaneously at the same frequency and acceleration).

2.3.2 Rectifiers

A rectifier is a kind of power circuit able to convert an incoming AC signal into a non-regulated output DC signal. These kinds of circuits are absolutely mandatory in those applications where the main power supply generates an AC power signal – like piezoelectric generators or inductive links- and the electronic system requires a DC power supply.

A diode is typically used as a main device for the classical half and full-wave rectification topologies; Fig. 2.17. In a half-wave configuration, diode D_1 rectifies

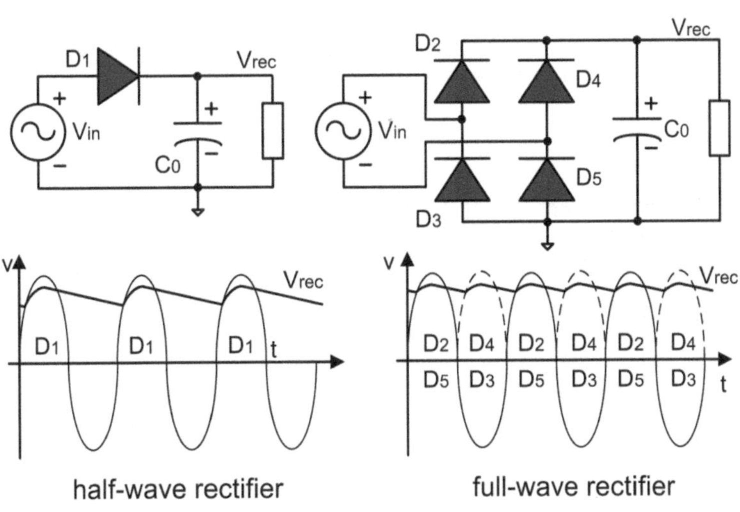

Fig. 2.17 Diode half-wave and full-wave rectifiers

only the positive part of the input voltage waveform while charging capacitor C_O. On the other hand, D_1 does not work during the negative period of the input signal and the load discharges C_O [53, 54]. The operating principle for the full-wave rectification is similar to the half-wave where diodes D_2 and D_5 rectify the positive part of the input wave and the negative part is rectified by D_3 and D_4.

These diodes topologies are widely used in medium and high power applications where the input AC voltage is much higher than the forward voltage of the diode (V_F). In those applications, V_F can be rejected and the efficiency of the rectifier is still high.

Both diode bridge rectification topologies – half and full-wave – becomes unsuitable methods for low-power low-voltage applications, where the main power generates a small voltage that is comparable with the forward voltage (V_F) of the diode, typically between 0.4 V for Schottky diodes and 0.6 V for p-n diodes. In these cases, V_F cannot be neglected and, consequently, low efficiency and rectified voltage are obtained. Moreover, the integration (design) of a diode rectifier in an integrated circuit (IC) is limited. First, sometimes the devices provided by the technology are restricted and diodes are not offered or their performances are not good enough to design a rectifier; secondly, IC integration brings the opportunity to develop more complex power conversion circuits with better performance for Low-Voltage Low-Power conversions. Following the idea of integrated rectifiers, three different types are presented. Two of them are based on the idea of synchronous rectifiers and the third one is a modified full-wave architecture formed by MOSFET transistors.

2.3.2.1 Synchronous Rectifier

A Synchronous Rectifier (SR) is composed by two main devices, one MOS transistor, working as a switch, and one Driving Amplifier (DA). A simple SR is presented in Fig. 2.18. The source-drain voltage across the N-channel MOSFET is sensed continuously by the amplifier. When the source voltage is positive with respect to its drain voltage the NMOS transistor conducts.

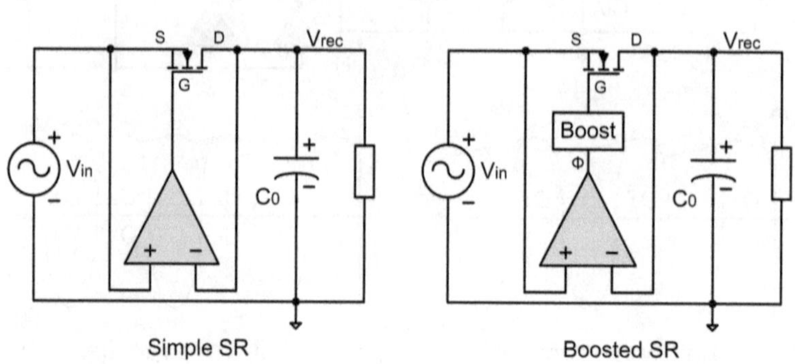

Fig. 2.18 Simple and boosted synchronous rectifiers (SR)

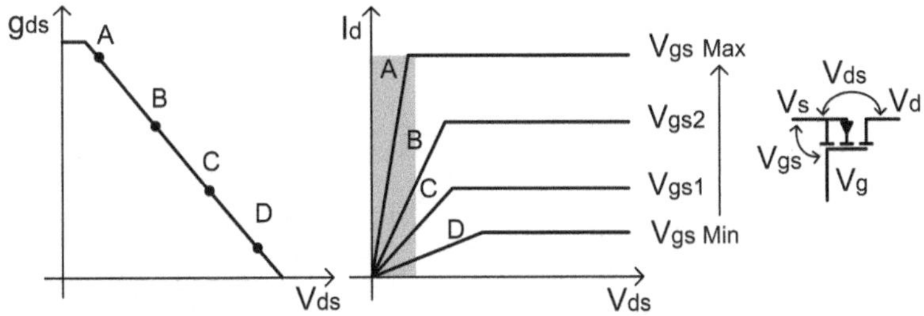

Fig. 2.19 Nmos g_{ds} behavior

The efficiency of the simple rectifiers depends on the voltage drop across the NMOS, the higher the drain-source voltage (V_{ds}), the lower the efficiency of the rectifier. Ideally, an NMOS transistor working as a switch acts as a fixed conductance – or fixed small resistance -. But, in practice, the drain-source conductance, gds, varies strongly with the input voltage signal (V_{in}) at the source of the NMOS. This effect is even more visible in low voltage applications, when the supply voltage is comparable with the threshold voltage of the transistors.

The behavior of gds in the linear zone is described in Eqs. (2.10), (2.11), (2.12) and Fig. 2.19. From Eq. (2.12) can be derived that the transistor works in the triode region involving a low conduction for input voltages (V_{in}) closed to V_g –waveform D-. On the other hand, when the input voltage decreases, the conductance increases due to the increase of the gate–source voltage (V_{gs}) difference – waveforms B and C-. Finally, with the maximum gate – source voltage difference, the highest conductance value is achieved – waveform A -. This maximum gate – source voltage value is defined by the used technology.

An extra boost stage has to be added to the simple SR, between the transistor and the amplifier Fig. 2.18, to reduce the on- resistance and to extend the linear range of the transistor, gray zone Fig. 2.19. This stage generates a gate voltage (V_g) higher than the supply voltage, without exceeding the maximum gate-source (V_{gs}) potential, to drive properly the gate of the transistor while increasing the g_{ds}.

$$g_{ds} = \frac{\partial I_d}{\partial V_{ds}} = \mu C_{ox}\frac{W_n}{L_n}(V_{gs} - V_{TH}^n - V_{ds}) \tag{2.10}$$

$$g_{ds} = \mu C_{ox}\frac{W_n}{L_n}(V_g - V_s - V_{TH}^n - V_{ds}) \tag{2.11}$$

$$g_{ds} = \frac{\partial I_d}{\partial V_{ds}} = \mu C_{ox}\frac{W_n}{L_n}(V_g - V_{in} - V_{TH}^n - V_{ds}) \tag{2.12}$$

The boost stage is based on the Bootstrap circuit [61, 62] depicted on Fig. 2.20. It turns the switch on and off using the signal Φ from the amplifier. When Φ is low, off phase, the switch gate is grounded across transistors M_7 and M_{10}. Meanwhile, the battery capacitor C_3 is charged to V_{CC} by M_3 and M_{12}. Switch and capacitor are isolated by transistors M8 and M_9. During the on phase, Φ is high, device M_8 is

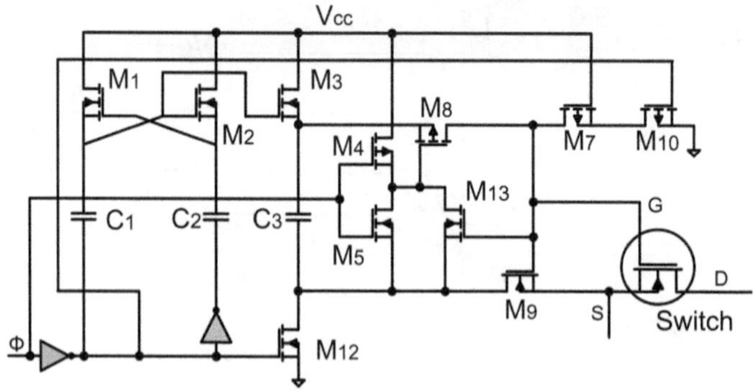

Fig. 2.20 Bootstrap architecture

activated by M_5 and the load flows from battery capacitor C_3 to the gate, activating it and M_9. Furthermore, transistor M_9 helps the Switch gate to keep the voltage at the source (S) shifted to V_{CC}. Devices M_1, M_2, C_1 and C_2 are not necessary for the circuit but ensure a unidirectional charge of C_3 by M_3.

Figure 2.21 depicts the simulated waveforms for both synchronous rectifiers (b and c) and for a half-wave diode rectifier (a). The diode rectification scheme (a)

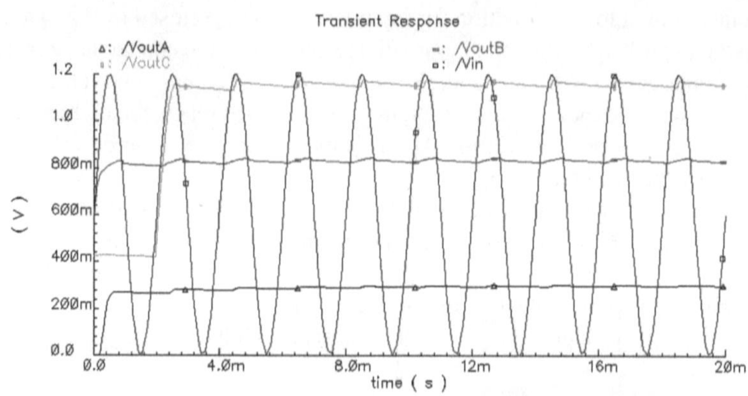

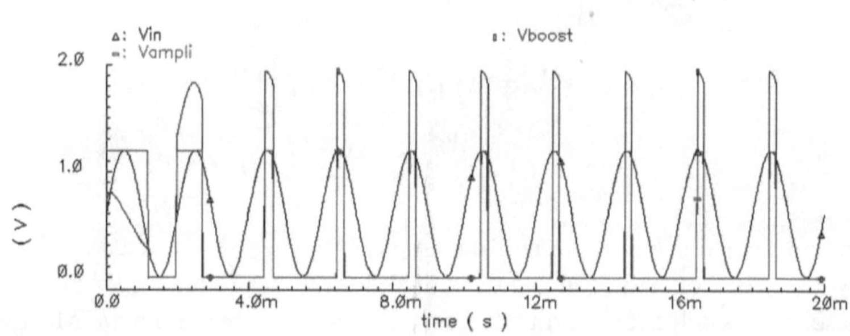

Fig. 2.21 *Top*: **a** half bridge **b** Simple SR and **c** Boosted SR simulation results. *Bottom*: Bootstrap circuit operation, **d** control signal Φ and **e** Φ shifted to V_{cc}. $C_O = 10\ \mu F$, $R_{Load} = 10\ k\Omega$ and $V_{in} = 1.2$ V @ 500 Hz

involves an enormous voltage drop due to the diode forward voltage. Its conversion efficiency is small –24.6%- because of the rectified DC value is around 300 mV. On the other hand, the schemes based on synchronous rectification increase considerably the rectified DC value, hence the conversion efficiency rises. Waveform (b) shows the output V_{rec} voltage value –833.5 mV- obtained with the Simple SR and (c) shows the output DC voltage value –1.16 V- obtained with the Boosted SR.

The maximum rectified DC voltage with both SR, Simple and Boosted is 282% and 393% respectively higher than the obtained with the diode rectification. At the bottom of the Fig. 2.21 the bootstrapped circuit operation in detail is presented. The control signal (Φ), generated by the amplifier (d), is shifted to V_{cc} (e), but, this voltage never exceeds the maximum V_{ds} defined by the technology. All the simulations has been carried out using a $C_0 = 10\ \mu F$, $R_{Load} = 10\ k\Omega$ and $V_{in} = 1.2\ V$ @ 500 Hz.

Both synchronous rectifiers are designed and experimentally validated in 130 nm CMOS technology [63, 64, 65]. The layouts are presented in Fig. 2.22. Figure 2.23 presents the start-up and steady state for both rectifiers for load conditions of

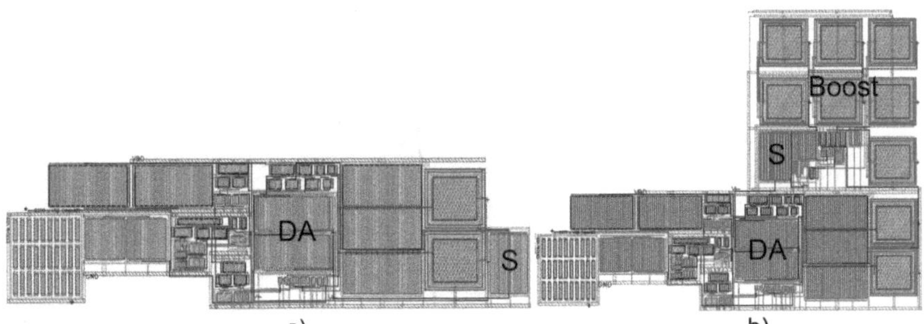

Fig. 2.22 **a** Simple (184 μm × 51 μm) **b** Boosted (170 μm × 127 μm) SR .DA = Driving amplifier, S =NMOS Switch

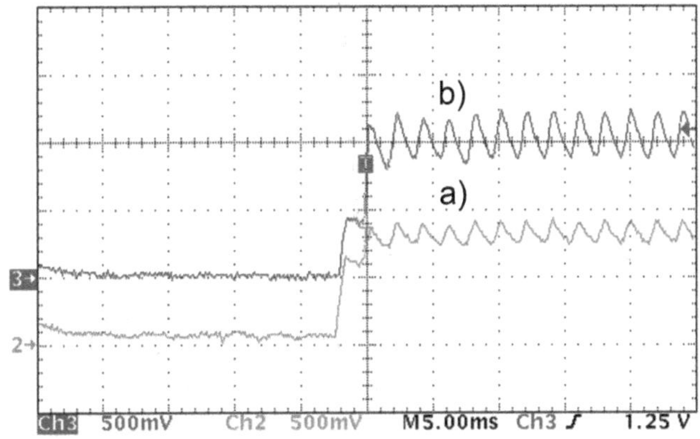

Fig. 2.23 Simple (**a**) and Boosted (**b**) synchronous rectifiers experimental results

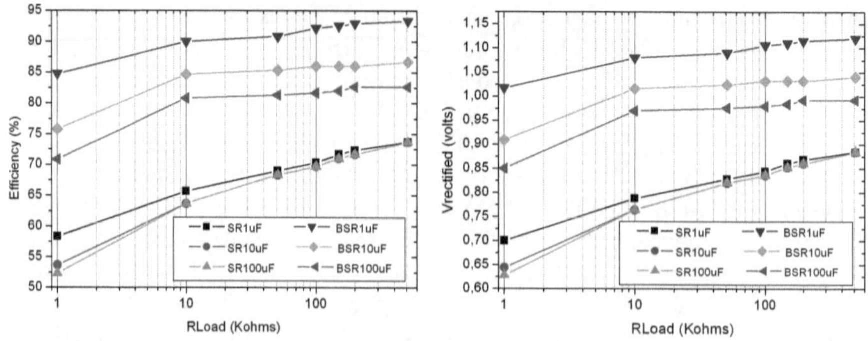

Fig. 2.24 Rectified voltage and efficiency for simple (SR) and boosted (BSR) rectifiers for several load conditions

10 kΩ / 10 μF. Figure 2.24 presents the rectified voltage and efficiency vs. several load conditions using three different load capacitors. The efficiency of the Boosted Rectifier (BSR) increases rapidly using higher load capacitors, whereas in the Simple one (SR) the efficiency keeps almost constant. Finally, Table 2.8 compares the most significant simulated and experimental values for both synchronous rectifiers.

The main drawback of these rectifiers concerns the use of an external power source to supply the driving amplifier (DA), i.e. a watch battery. A key aspect for the future is to use the same input AC signal to power the amplifier in order to obtain a full autonomous system.

Table 2.8 Simple and boosted synchronous rectifiers maximum characteristics

Load:	Simulated		Experimental	
600 k Ω / 10 μ F	Simple SR	Boosted SR	Simple SR	Boosted SR
V_{rec}	833.5 mV	1.16 V	847 mV	1.11 V
P_{rec}	69.5 μW	134.6 μW	70.8 μW	122 μW
Efficiency (η)	69.4%	96.6%	70.6%	91.1%

Driving Amplifier (DA)

A class AB operational amplifier, Fig. 2.25, has been selected to reduce the power consumption and to be able to drive enough current to drive the NMOS switch. Figure 2.26 depicts its schematic (e) [44, 66]. The biasing circuits (b, c and d) and the 1 μA self-biased current source (a) are also presented. Table 2.9 presents the main features.

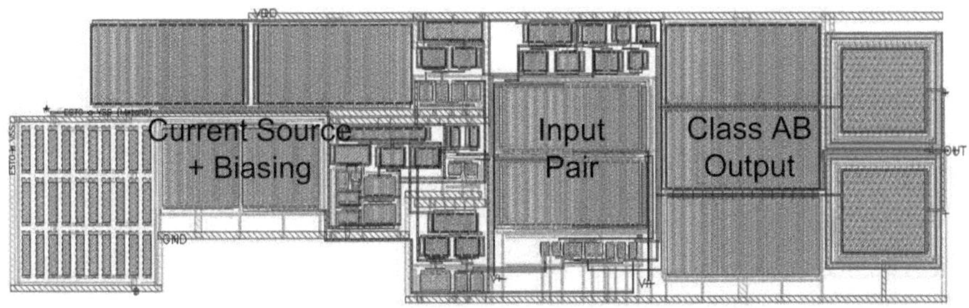

Fig. 2.25 Class AB operational amplifier layout (170 μm × 51 μm)

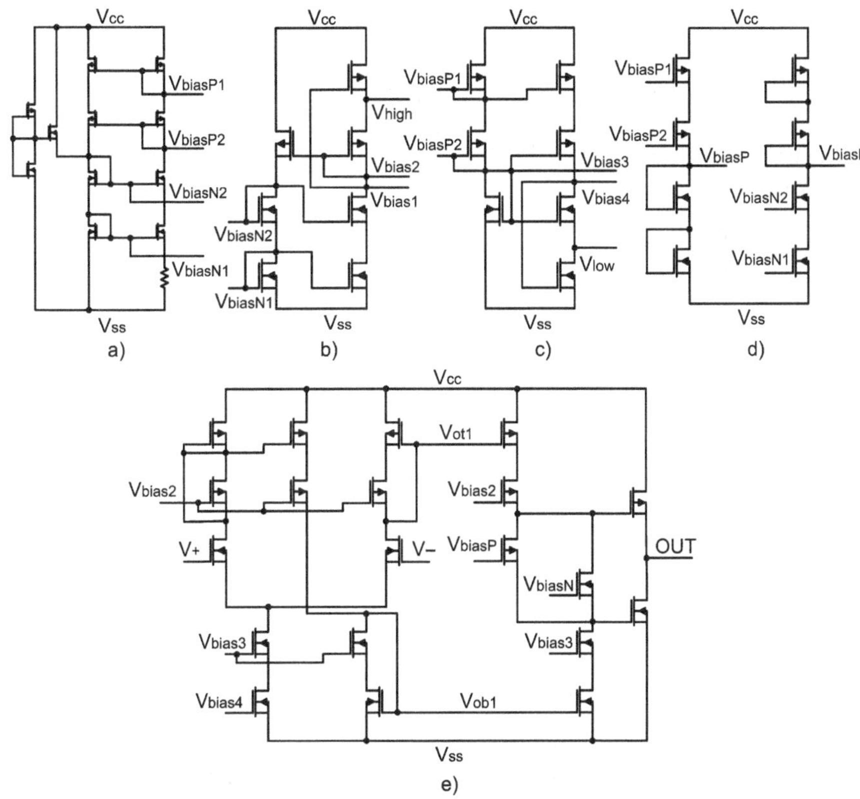

Fig. 2.26 Class AB operational amplifier (**e**) and its biasing circuits (**b**, **c**, **d**) and current source (**a**) designed as a driving amplifier (DA)

2.3.2.2 Full-wave MOSFET Rectifier

The NMOS and PMOS transistors are widely used to design integrated half and full-wave AC/DC rectifiers instead of diodes, with an enormous drop voltage for

Table 2.9 Class AB operational amplifier electrical characteristics ($V_{CC} = 1.2$ V)

Symbol	Parameter	Value	Units
G_{DC}	DC gain	65	dB
F_U	Unity frequency	145 k	Hz
V_{Offset}	Offset voltage	2.4 m	V
PSSR+	V_{CC} power supply rejection ratio	62	dB
PSSR−	V_{SS} power supply rejection ratio	70	dB
SR	Slew rate	0.005	V/μs
V_{IN}	Nominal input voltage	1.2 (up to 2.5)	V
I_{CC}	Current supply @ 1.2 V	11 μ	A
P_{CC}	Power consumption @ 1.2 V	13 μ	W

low-voltage applications. There are several approaches to accomplish the rectification like those presented in [67, 68].

This section introduces the design of three full-wave rectifiers, one based on a NMOS and the other two based on PMOS transistors. A theoretical study is carried out for all rectifiers and an experimental validation for the NMOS and one PMOS rectifiers is then presented.

NMOS Full-wave Rectifiers

A bridge rectifier based on NMOS transistors [69] is presented in Fig. 2.27a. This topology presents a lower drop voltage during the rectification because the transistors start to conduct the current when the input voltage is close to the threshold voltage of the transistor, which is lower than the diode forward voltage.

The voltage is rectified as follows: when the input voltage polarity is positive the current flows through transistors M_1 and M_4 and transistors 2 and 3 are switched off. When the polarity is reversed, $V_{in}-$ is higher than $V_{in}+$, the current flows through transistors M_2 and M_3 while M_1 and M_4 are shut off.

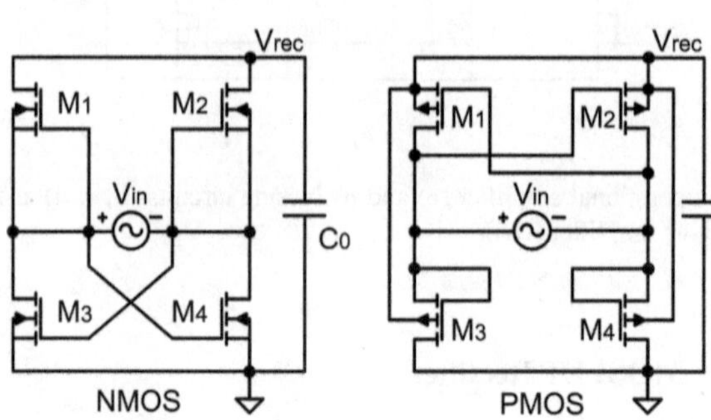

Fig. 2.27 NMOS and PMOS full-wave rectifiers

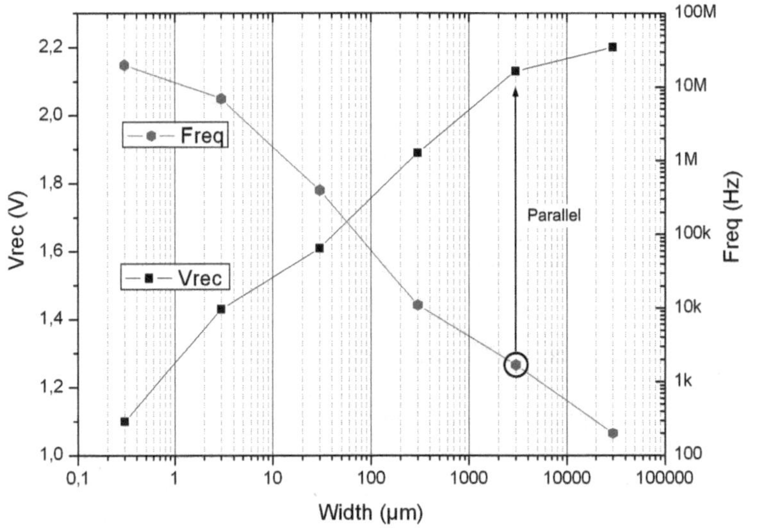

Fig. 2.28 Rectified voltage (V_{rec}) and operation frequency in function of transistor width for the NMOS full-wave rectifier of Fig. 2.27a (L= 1 um, $V_{in} = 5\ V_{pp}$ @ 500 Hz)

When working with MOSFET rectifiers a trade-off in terms of efficiency and area has to be analyzed. A low drop voltage across the transistors implies high channel width and low frequencies operation; on the other hand, small channel width represents a higher drop voltage and high operation frequencies. Figure 2.28 depicts the rectified voltage (V_{rec}) and the operation frequency (Freq) for the NMOS rectifier in Fig. 2.27a as a function of a single NMOS transistor width. A sinusoidal 5 V @ 500 Hz signal is used as input.

When the input voltage value is low and at low frequencies, like when working with a PZT, it is a good option to select high width transistors with low drop voltages. On the other hand, when working at higher frequencies, like in inductive powering, small widths are a good choice in order to increase the switching frequency. There is an optimal operation point in the intersection of both waveforms but, if the desired working frequency and the expected rectified voltage are higher than those obtained in the optimal point, some extra considerations have to be done.

Several transistors in parallel can be connected to improve the performance of the rectifier in terms of voltage drop and frequency [70]. This configuration divides the current into all transistors reducing the drop voltage (V_{ds}) of each parallel NMOS. Hence, it is possible to improve the working frequency because the size of all individual transistors is smaller than just one big transistor. Figure 2.29 presents the conception for two NMOS in parallel.

A single transistor with a size of W is able to conduct a drain-source current I_{d1} involving a drop voltage of V_{ds1}. When two transistors with identical width (W) are connected in parallel, the drop voltage is reduced to V_{ds2} because the current

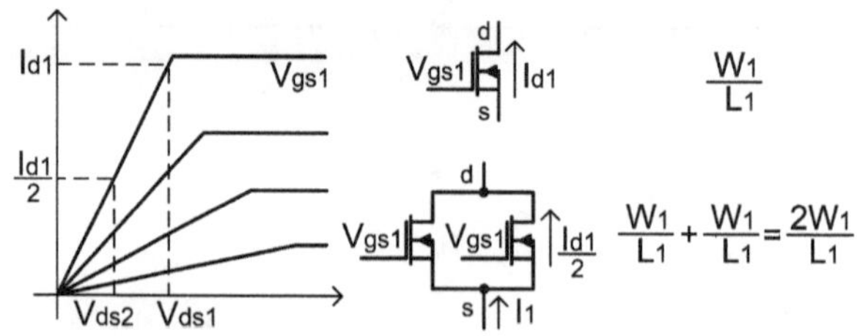

Fig. 2.29 Effect of MOS transistors parallel connection

flowing through each one is divided by two and the effective width for the parallel transistor is 2 W.

Taking into account the previous considerations, an integrated NMOS rectifier (Fig. 2.27a) formed by four transistors – M_1, M_2, M_3 and M_4 – of 3,000 μm width is implemented. This size involves a voltage drop range of 0.2–0.6 V depending on the technological drifts. To solve the low working frequency due to the size, black circle Fig. 2.28, a parallel structure of 300 transistors of 10 μm is used to form each transistor. Hence, it is achieved a new working frequency range of 3–15 MHz, depending on the technological drifts.

Figure 2.30 depicts the final layout of the NMOS rectifier presented in Fig. 2.27a. The rectifier is symmetrically in both axis, X and Y, and an interdigitated common-centroid structure is selected to integrate all four transistors. In that way, problems regarding layout matching between transistors and leakage currents are minimized.

The layout is divided in two main parts: the high part composed by transistors M_1 and M_2 and the lower part with M_3 and M_4. All of them are interdigitated. Each transistor is formed by 30 functional blocks (A, B, C or D) and each functional

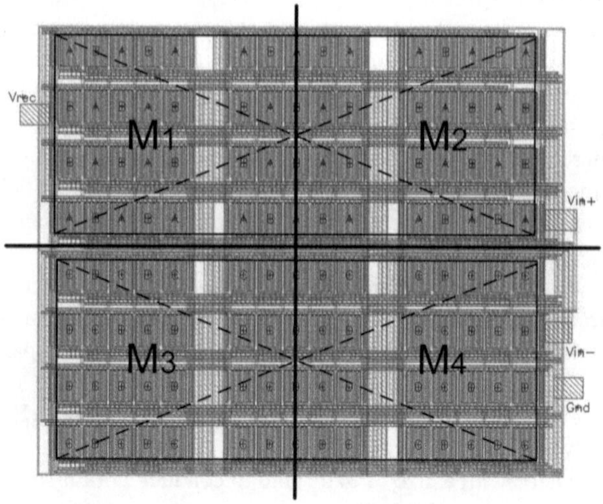

Fig. 2.30 NMOS rectifier layout (180 μm × 150 μm)

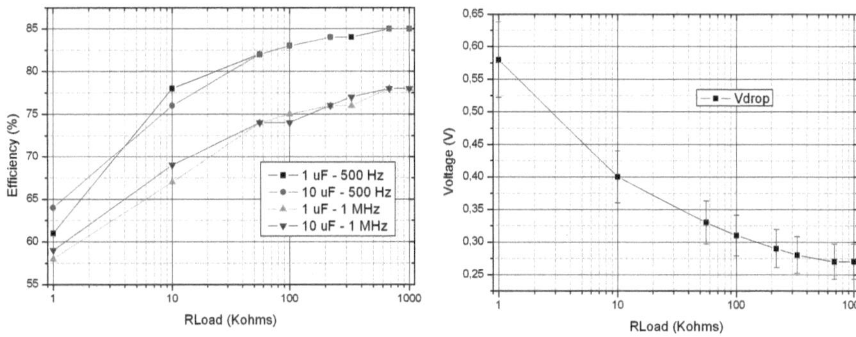

Fig. 2.31 NMOS rectifier efficiency and voltage drop for several load conditions

block is formed by 10 transistors with 10 μm of width. Transistor M_1 is formed by $M_1 = 30 \text{ x } A = 30 \times 10 \times 10\ \mu m = 3{,}000\ \mu m$. The other transistors are designed following the same procedure respectively.

An efficiency range of 55–85% is achieved depending on the load and frequency conditions. In the case of high load the efficiency decreases due to the voltage drop across the transistors. On the other hand, a better efficiency is obtained in the case of low current loads: lower the current flowing through the rectifier lower the drop voltage through it, hence higher is the efficiency.

The operation frequency also modifies the efficiency of the rectifier. For higher frequencies the transistor leakage increases producing more losses so, the efficiency decreases. Figure 2.31 depicts the efficiency of the rectifier vs. several load conditions for different frequencies and filter capacitances (C_0) and the voltage drop across the rectifier is function of load conditions. As can be noticed, the efficiency of the NMOS rectifier is 10% lower than the obtained with the boosted synchronous rectifier (Section 2.3.2.1) for the same load conditions.

One of the pitfalls of the rectifier is the dependence of the efficiency with the transistor voltage drop. A trade off in terms of area, load current, leakage and voltage drop has to be reached to fit the correct rectifier in each application. Also, a minimum input voltage higher than the threshold voltage of the transistors is necessary to activate them properly and rectify the incoming signal.

The use of an external battery is not necessary in the NMOS passive rectifiers. It represents a great advantage compared to the Synchronous Rectifiers to develop a Self-Powered application. Moreover, these rectifiers can be easily integrated and adapted to any design and no extra components like boost stages or driving amplifiers are necessaries.

On the other hand, a good layout is important to avoid matching problems. Table 2.10 summarizes the main features of the rectifier and Fig. 2.32 presents the minimum input voltage that can be properly rectified by the NMOS AC/DC converter. It is also important to take into account the temperature dependence of the leakage current in those applications designed to work in a wide range of temperatures.

Table 2.10 NMOS Full-wave rectifier electrical characteristics

Symbol	Parameter	Min	Max	Units
V_{IN}	Nominal input voltage	0.25	2.5	V
V_{drop}	Drop voltage	0.05	0.6	V
I_{out}	Output current	–	20	mA
Freq	Working frequency	–	16	MHz
η	Efficiency	52	85	%
$I_{leakage}$	Leakage current	–	1.4	mA
W	Width of each transistor	3,000		μm
L	Length of each transistor	0.28		μm

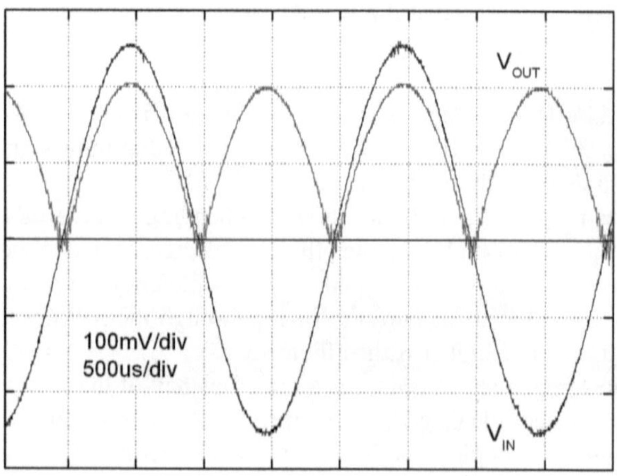

Fig. 2.32 NMOS rectifier minimum input voltage at no-load condition

PMOS Full-wave Rectifiers (Bridge and Voltage Doubling)

Figure 2.27b depicts the architecture of the PMOS bridge rectifier. When voltage $V_{in}+$ is higher than $V_{in}-$ transistors M_2 and M_3 are activated whereas M_1 and M_4 are shut off. On the other hand, when voltage $V_{in}-$ is higher than $V_{in}+$ transistors M_2 and M_3 are deactivated and M_1 and M_4 are turned on [68, 71, 72].

The drop voltage of the PMOS transistors is controlled and minimized by selecting properly the size of each one. The higher the PMOS size, the lower the drop voltage and frequency switching. The steady state leakage current is also minimized connecting the bulk to the highest potential.

The PMOS rectifier presents a simulated efficiency of 72% under the same conditions as the NMOS, Fig. 2.34. In spite of its better efficiency the NMOS rectifier is integrated due to its reduced dimensions, the PMOS rectifier area is 10 times higher than the NMOS.

The previous NMOS and PMOS rectifiers generate a positive DC voltage. Sometimes, a symmetrically voltage supply is necessary for all those applications that requires a positive and negative DC voltages to power the circuitry. Figure 2.33

Fig. 2.33 Full-wave voltage doubling rectifier and its minimum input voltage

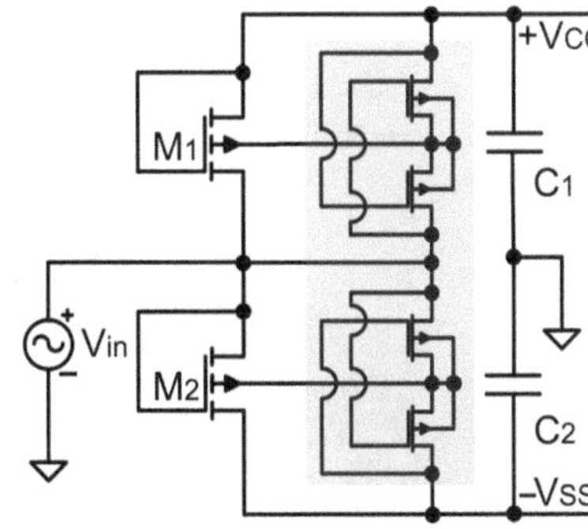

depicts a full-wave voltage doubling analyzed and implemented [53]. Both transistors are diode connected. During the positive cycle the current flows through transistor M_1 generating the positive DC voltage – V_{CC} – whereas M_2 conducts the current during the negative period of the input signal producing the negative DC voltage – V_{SS}-.

The bulk of each transistor is always connected to the highest potential using an auxiliary circuit (grey zone Fig. 2.33) to avoid a forward biasing and to guarantee the lowest current consumption through the bulk [62, 73]. A small size is used for those auxiliary transistors (W = 30 μm, L = 0.35 μm). Otherwise the leakage produced by these transistors becomes too large at high frequencies.

Transistors M_1 and M_2 has been designed following the same considerations done in the previous NMOS rectifier. The selected width of the PMOS transistors is 3,000 μm using 300 transistors of 10 μm in parallel. The final layout has the same appearance as transistors 1 and 2 in Fig. 2.30 with a new dimensions of 180 μm × 80 μm. Table 2.11 summarizes the main characteristics. It can be noticed that all the values are lower than the ones obtained with the NMOS rectifier.

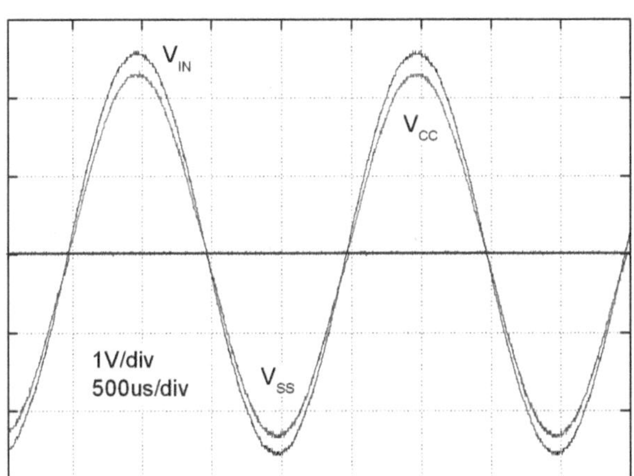

Fig. 2.34 PMOS rectifier minimum input voltage at no-load conditions

Table 2.11 Full-wave voltage doubling rectifier electrical characteristics

Symbol	Parameter	Min	Max	Units
V_{IN}	Nominal input voltage	±0.3	±2.5	V
V_{drop}	Drop voltage	0.1	0.82	V
I_{out}	Output current	–	17	mA
Freq	Working frequency	–	14	MHz
η	Efficiency	44	72	%
$I_{leakage}$	Leakage current	–	1.14	mA
W	Width of each transistor	3,000		μm
L	Length h of each transistor	0.28		μm

2.3.3 Regulation

As it is previously commented, this module regulates the DC voltage provided by the rectification stage. There are several options to regulate and to control the voltage after AC/DC rectification like DC/DC converters [74], Zener diodes or linear regulators [75]. One of the most used circuits for low voltage applications, where the generator produces low voltage, is based on LDO's (Low Dropout Regulator). A 1.2 V LDO is designed and explored in Section 2.2.2 and it can also be used after the rectification module as long as the rectified voltage overcomes the minimum input voltage of 1.3 V.

Two extra regulations methods are explored in this section. The first one uses a Charge Pump (CP) to push the input voltage to a higher value, and the other method is designed specifically for self-powered low voltage low power applications.

2.3.3.1 Charge Pump (CP)

A charge pump (CP) is a type of DC/DC converter able to pump the input voltage to a higher value avoiding the use of inductors. Its purpose basically consists on generating a voltage level two times higher than the input voltage ($V_{out} = 2\ V_{in}$). The adopted charge pump option is based on a 2-phase scheme [76, 77, 78]. Also a higher voltage could be obtained adding more stages [79].

This 2-phase option is good enough in terms of area and power consumption, instead of solutions based on complex architectures like in [80, 81, 82]. An option based on an n-stage Dickson charge pumps is dismissed because of the necessary area for the layout implementation and the very poor efficiency due to the losses in the diodes. Furthermore, CP architectures based on four clocking schemes and solutions that switch the bulk connections to the highest voltage are also dismissed for their complexity and power consumption; they use several drivers, level shifters and extra electronics.

The implemented TPVD (Two Phase Voltage Doubler) charge pump is depicted in Fig. 2.35a. The four switches are turned on and off periodically in two modes of operation: a store mode and a transfer mode. In the store mode the switches M_0 and M_2 are switched on, while the other two switches are off. The C_{fly} capacitor is then charged to the V_{in} voltage. In the transfer mode, switches M_0 and M2 are OFF and

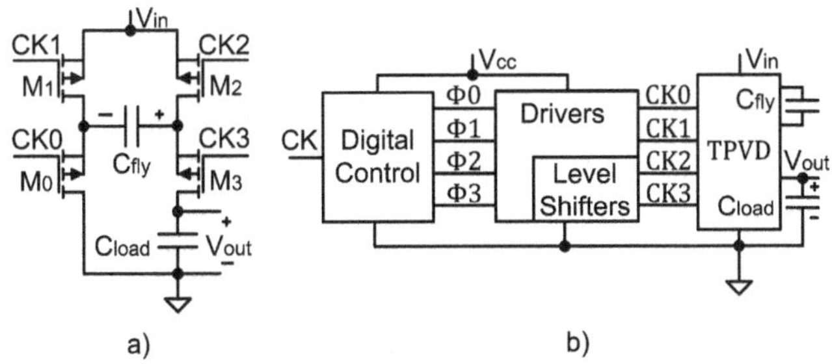

Fig. 2.35 **a** TPVD schematic approach and **b** TPVD + auxiliary circuits

switches M1 and M3 are on. Theoretically, if there are no losses, the TPVD doubles the V_{in} voltage at the unregulated output node, V_{out}.

The charge pump uses a simple digital core that generates the suitable control signals (low voltage) to drive the four switches. These signals are generated from a signal clock circuitry. In order to work with the PMOS transistors, a level shifter driver is introduced to drive the gate of transistors M_2 and M_3 to the maximum available voltage in the charge pump, V_{out}, in order to define the OFF condition of these transistors. In Fig. 2.35b the TPVD with the auxiliary circuits is depicted.

The digital control generates the signals in such a way no-crossing conditions exist. Crossing conditions have to be avoided, for instance, between the On-Off transitions of transistors M_2 and M_3, and also between M_0 and M_1 to prevent short circuits between V_{in} and GND. A dynamic level shifter [83] is designed to drive the PMOS transistors M_2 and M_3.

The designed TPVD layout is presented in Fig. 2.36. Some of the most interesting simulations are presented in the following. Figure 2.37 depicts the start-up and steady state operation of the charge pump at $V_{in} = 1.2$ V, CK = 1 MHz and Load = 100 kΩ. It is also possible to observe the clock signals used to drive the transistors and how clocks 2 and 3 are shifted to the output voltage, maximum available

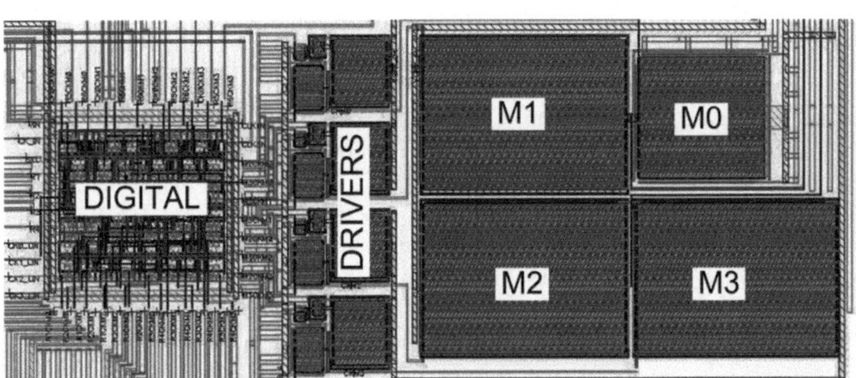

Fig. 2.36 Layout of the implemented TPVD (230 μm × 195 μm)

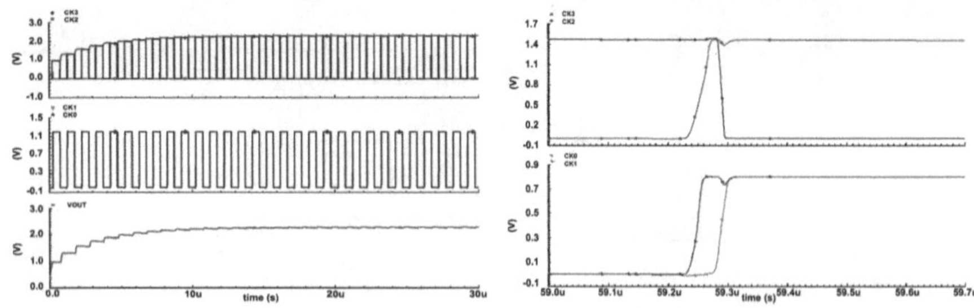

Fig. 2.37 TPVD simulated start-up and steady state operation (*Left*) and non-overlapped clock signals (*Right*). $V_{in} = 1.2$ V, CK = 1 MHz and Load = 100 kΩ

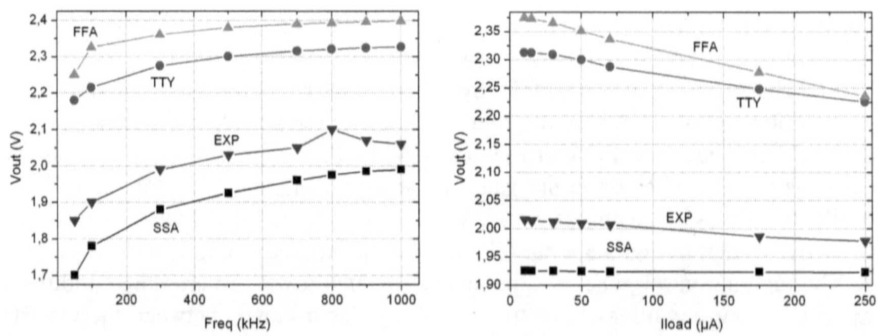

Fig. 2.38 Simulated and experimental TPVD output voltage in function of switching frequency (*Left*) at no load and vs. several load conditions (*Right*) (500 k Hz.; $V_{in} = 1.2$ V)

voltage, to drive transistors M_2 and M_3. It is also presented in detail the dead times between the clock signals and the no-crossing conditions between them.

Figure 2.38 presents the simulated output voltage (V_{out}) for $C_{fly} = C_{load} = 3.3$ nF in open load condition vs. clock frequencies for different statistical models, SSA (Slow case), TTY (Typical case) and the FFA (Fast case). It is also depicted the V_{out} value for $C_{fly} = C_{load} = 3.3$ nF vs. different load conditions @ 500 k Hz ($V_{in} =$ 1.2 V).

A set of measurements are carried out to the IC in order to check the performance of the unregulated charge pump. The experimental output voltage drops from the ideal theoretical value of 2.4–2.0 V. The behavior of the implemented TPVD follows the SSA simulated model case. For instance, in the condition of no-load, $C_{fly} =$ $C_{load} = 3.3$ nF and clock of 500 kHz the experimental V_{out} is 2.035 V and the simulated SSA value is around 1.92 V. Figure 2.38 presents the experimental output voltages, V_{outEXP}, which can be compared with the simulated results.

An output voltage around 1.9 V is obtained for frequencies up to 200 kHz. Increasing the switching frequency, the output series resistance of the charge pump decreases [84], and the output voltage increases up to 2.10 V for an optimal frequency of 800 kHz.

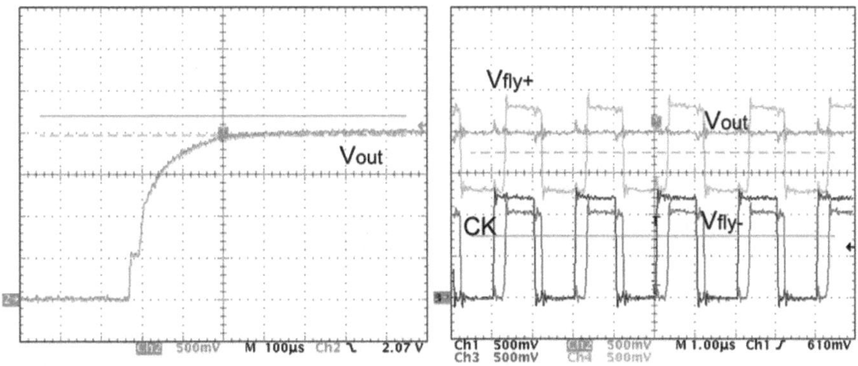

Fig. 2.39 Experimental start-up (*Left*) and flying capacitor voltages (*Right*) (load of 3.3 nF and 100 kΩ, C_{fly} of 3.3 nF, clock of 100 k Hz and $V_{in} = 1.2$ V)

Figure 2.39 presents the experimental start-up with a load of 3.3 nF and 100 kΩ, C_{fly} of 3.3 nF, clock of 100 kHz and $V_{in} = 1.2$ V. Increasing the clock frequency, lowers the start-up time and the output ripple. The same figure depicts the experimental voltages at different nodes of the charge pump. Capacitor C_{fly+} changes its value from V_{in} to 2 V_{in} when the charge pump phase changes from store mode to transfer mode, whereas C_{fly-} changes from 0 to V_{in}. It is also depicted the drop voltage defined as 2 $V_{in} - V_{out}$.

Figure 2.40 presents the output voltage (V_{out}) at open load condition, for different switching frequencies and different values of C_{fly} and C_{load} capacitors. C_{fly} capacitor should be adjusted to boost the output voltage taking into account the desired power level. This capacitor should be higher than the parasitic capacitances in the nodes. From simulations a value of 3.3 nF was selected and the experimental results confirm a better performance of the charge pump using that value. As the value of the C_{fly} capacitor decreases, the performance of the charge pump is even worse.

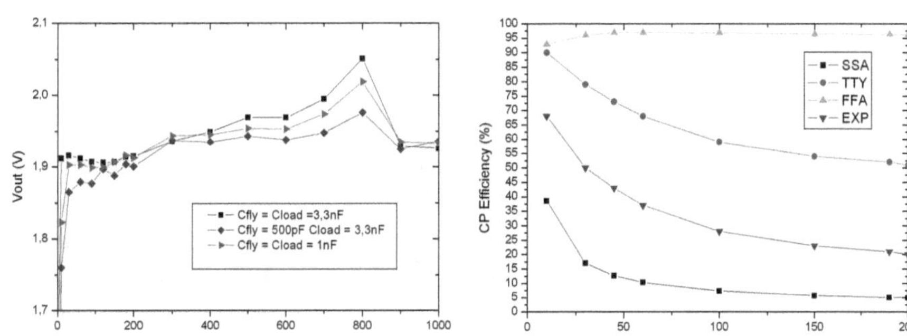

Fig. 2.40 Experimental output voltage vs. frequency switching for several C_{fly} and C_{load} capacitors (*Left*) and experimental efficiency for the particular case of $C_{fly} = C_{load} = 3.3$ nF @ 800 kHz and Vin = 1.2 V

Table 2.12 TPVD electrical characteristics

Symbol	Parameter	Min	Max	Units
V_{in}	Nominal input voltage	0.4	1.2	V
I_{out}	Output current	–	250 μ	A
η	Efficiency	19	68	%
Freq	Working frequency	100 k	1 M	Hz
V_{DROP}	$2\,V_{in} - V_{out}$	0.1	0.5	V
V_{CC}	Supply voltage	1	1.2	V

This situation is defined by the value of the parasitic capacitors that are responsible of the switching losses, if they have a value comparable with C_{fly}, the features of the CP decrease.

The best experimental efficiency is presented in Fig. 2.40 for the particular case of $C_{fly} = C_{load} = 3.3$ nF @ 800 kHz. It is compared with the theoretical efficiencies obtained by simulation with the different models. As it is commented the IC is placed in the region of slow mobility, SSA. The main features of the designed TPVD are summarized in Table 2.12.

This kind of inductorless DC/DC converter is a good option to develop on-chip application because only few external capacitors are necessary. Furthermore, it is possible to work with low input voltages (> 1 V) and produce usable voltages (< 1 V) thanks to its pumping ability and the possibility to connect more than one stage in parallel [79]. In low power applications Charge pumps are widely used with thermoelectric [85], magnetically induced [86] and piezoelectric generators.

On the other hand, the main drawback of the TPVD Charge Pumps is related to the external supply voltage necessary to power the internal circuits. Moreover, an on-chip clock generator has to be included in the design increasing the power consumption.

This TPVD CP can be connected together with the Boosted Synchronous Rectifier (Section 2.3.2.1) to validate the generic architecture proposed in Fig. 2.14. The SR is able to work with a low AC input voltage and the DC/DC Charge pump generates an unregulated DC voltage two times higher than the rectified voltage. Figure 2.41 presents the connection schematic.

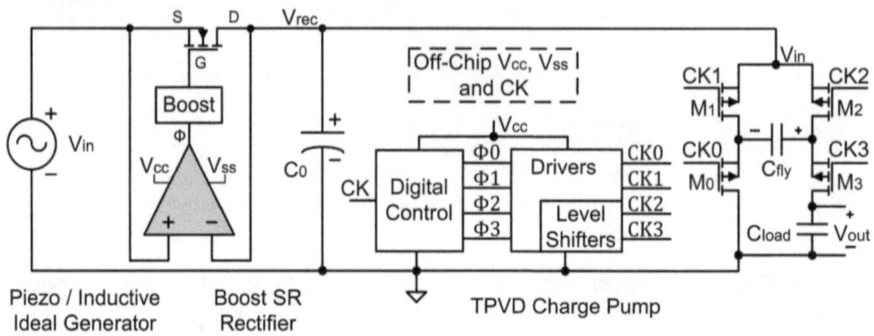

Fig. 2.41 Schematic of the boost synchronous rectifier and TPVD connection

Other types of Charge Pumps that combine rectification and pumping capabilities are also available. Some examples are found in [87, 99] where two CMOS UHF Rectifiers are used in RFID Tags.

2.3.3.2 Self-Powered Regulation Control Module (SPRCM)

In Low-Power applications the use of DC/DC converters or other type of classical regulations methods [74] are not possible because consumption represents a high percentage of the incoming power, hence, lower efficiency. Moreover, the integration of some of these converters and regulators becomes difficult in Self-Powered Micro Systems (SPMS) because of external batteries or supply systems are necessary. In those cases, specifically designed control circuitry has to be implemented.

The new circuit presented above is based on a control module able to manage the power generated by the low power source and deliver it to the load in a usable form. The module presents a low quiescent current to assure low power consumption, in that way, the major part of the input energy is transferred to the load and good efficiencies are achieved.

Furthermore, the objective is to avoid the use of extra batteries to power the extra circuitry and to work only with the power generator and one storage element, like a capacitor or a super capacitor [89, 90, 91], in order to define a SPMS. The module can be used with several low-power input generators although it is focused to work with AC generators like magnetically induced and especially piezoelectric generators.

The proposed IC architecture is presented in Fig. 2.42. It is composed by one integrated switch transistor, one integrated Control Module, one off-chip storage element, a capacitor or super capacitor (SCapa), and if necessary, some external resistances and an AC/DC rectifier if the input voltage (V_{in}) is not a DC signal.

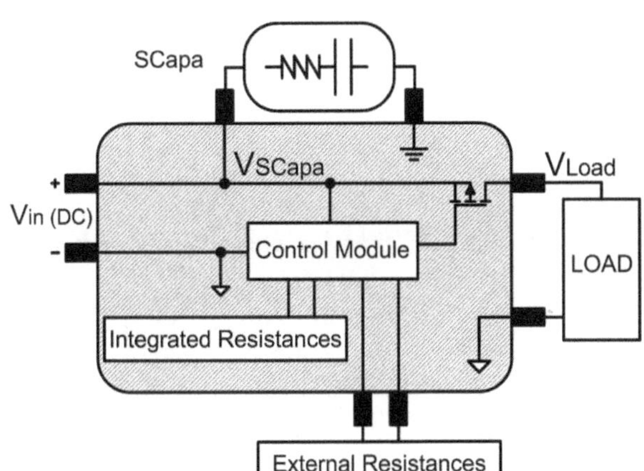

Fig. 2.42 Proposed architecture for self-powered regulation control module. Grey zone represents the integrated IC

The rectifier is used to convert the AC signal coming from the AC generator into an unregulated DC voltage. For those generators that directly provide a DC voltage this circuit is not necessary.

The unregulated DC voltage is used to charge the voltage at the super capacitor (V_{SCapa}). This voltage is sensed continuously by the Control Module and compared with two trigger values (V_{max} and V_{min}) in order to control the PMOS switch.

When the DC voltage at the SCapa reaches the maximum value (V_{max}) the Control Module switches on the PMOS transistor and the energy stored in the capacitor or battery is transferred to the load. At that point, V_{SCapa} starts to decrease (Discharging Phase) until a minimum voltage value (V_{min}).

When the Control Module senses the minimum threshold value the PMOS is automatically deactivated and the voltage transference from the capacitor to the load stops. Immediately, the voltage at the SCapa increases again up to V_{max} defining the Recharge Phase. It must be stated that the load must work between V_{max} and V_{min} voltages. The first time the battery is charged to V_{max} is defined as Start-Up phase. Figure 2.43 presents a simulation of the system with the three operation phases, V_{SCapa} and V_{Load}.

The time involved in the Start-Up and Recharge Phase varies in function of the power available by the generator and the value of the capacitor. The Discharge Phase time depends on the load power consumption. The energy is transferred to the load in pulses defining a Burst mode of operation: the load is only connected if there is enough energy at the capacitor.

It is possible to move from Burst mode toward Continuous mode if the load consumption is lower than the generated power. In that way, the load capacitor does

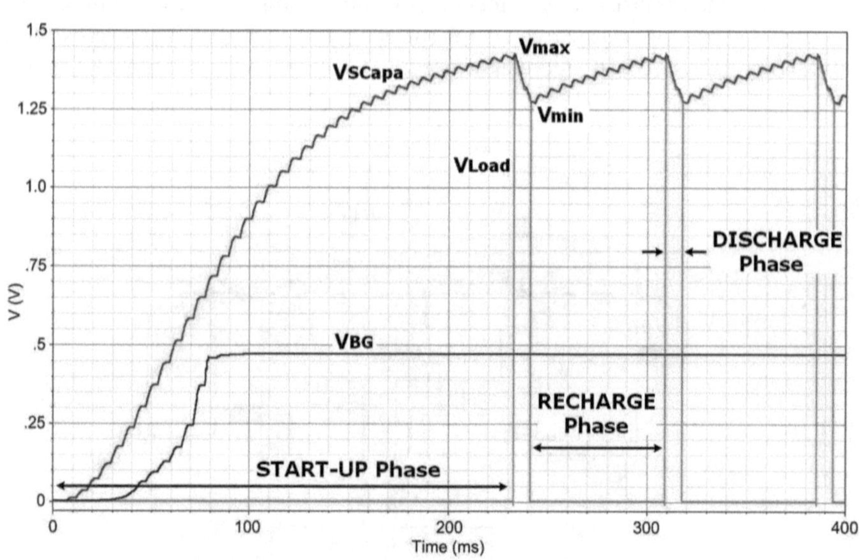

Fig. 2.43 Simulated performance of the module (SCapa = 47 μF)

not reach the minimum threshold voltage and a long operation discharge phase is defined.

Two different architectures have been used to implement the Control Module. One is based on a complex design with a precision band-gap reference circuit, comparators to define the max and min voltages, and a small digital logic.

The other architecture is based only in the use of a single comparator. The complex control has the ability to integrate extra functionalities like voltage thresholds programmability, Power-On Reset or sensors integration. On the other hand the simple control presents power consumption lower than the complex one. Both architectures are explained in detail in the following of this section.

Complex Control Module

The architecture is presented in Fig. 2.44. The system is formed by two low-power comparators (Fig. 2.45), which sense continuously the voltage at the storage capacitor (V_{SCapa}) and compare it with a reference voltage supplied by a Bandgap reference circuit, BG. Two resistive networks, R_{N1} and R_{N2}, are used to adapt the voltage level of V_{SCapa} to the range defined by the BG. One comparator is prepared to detect when the voltage reaches V_{max} and the other one is prepared to sense V_{min}.

The comparator is designed with 20 mV of hysteresis [70] to better define the voltage detection points and to avoid undesirable commutations, Fig. 2.46. In order

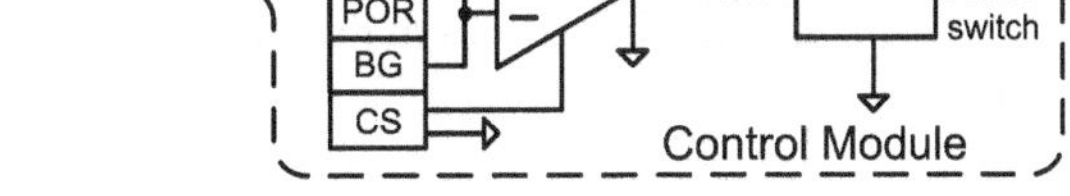

Fig. 2.44 Complex control module architecture

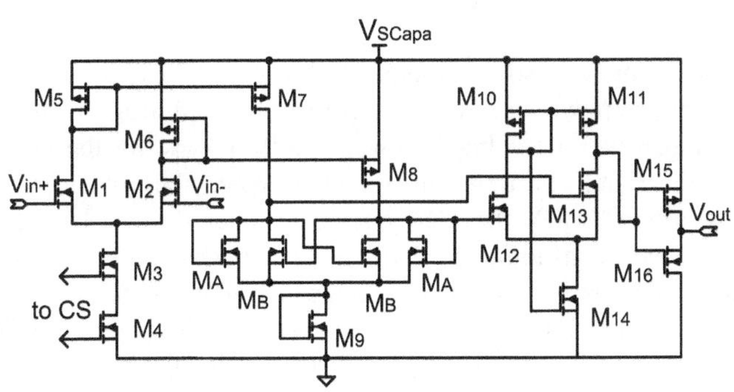

Fig. 2.45 Comparator architecture

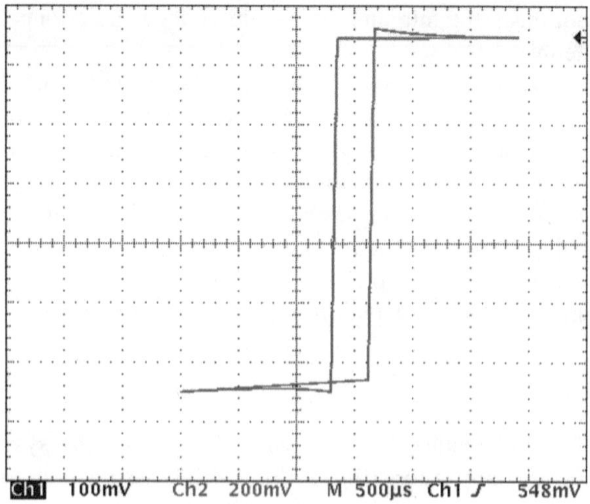

Fig. 2.46 Comparator 20 mV hysteresis @ 1.2 V

to reduce the power consumption both comparators share the same current source (CS) and Bandgap circuit (Section 2.2.2).

The signals generated by the comparators are used by a logic circuitry to generate the control signal for the PMOS switch and define the Charge and Discharge phases of the system. The comparators, the digital logic and the rest of the elements of the control unit are powered by the voltage at the storage element (V_{SCapa}).

As previously commented, the comparators are in charge of detecting the max and min voltages using a resistor network. In that way, several values of V_{max} and V_{min} can be programmed. To add an extra functionality in the final implementation, the control module can be configured to work with an on-chip resistance with a fixed detection values of $V_{max} = 1.1$ V and $V_{min} = 1.0$ V or with an external off-chip resistance allowing the system to detect voltages up to 2.3 V.

Figure 2.47 depicts the max and min values programmed to work between 1 V and 2 V using external resistors. The AC input voltage is rectified by the NMOS rectifier presented in Section 2.3.2.2. A piezoelectric transducer is used as a primary source.

The Complex Control also incorporates a Power-On Reset (POR). This circuit is used to reduce the power consumption of the module during the Start-up phase. At the beginning, when the voltage grows from 0 to V_{max} for the first time, only the POR and the Bandgap circuit are working. When the voltage at the storage capacitance is around 0.8 V the POR circuit activates the rest of the electronics increasing the power consumption of the module. In that way, the system is only connected when a minimum amount of voltage is available at the storage device.

Moreover, the Start-Up time is reduced because the system is deactivated. Figure 2.48 presents the architecture of the POR [92] and Table 2.13 compares the power consumption of the module during the start-up (0.5 V) using POR and without it.

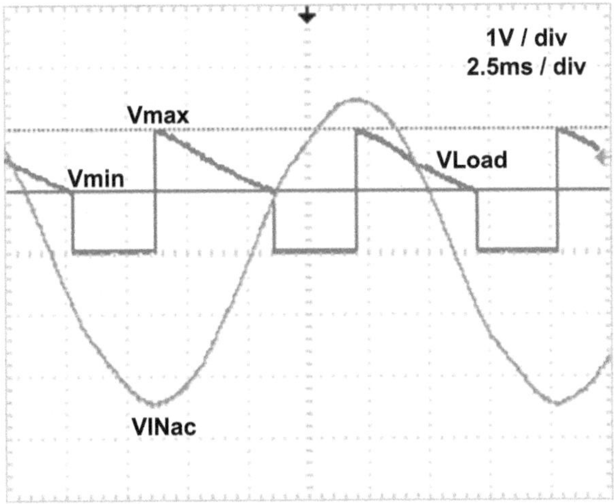

Fig. 2.47 Experimental output voltage using the Complex Control Module and an external network resistance. $V_{max} = 2$ V, $V_{min} = 1$ V and $V_{in} =$ AC of 2.5 V_{pp} at 50 Hz

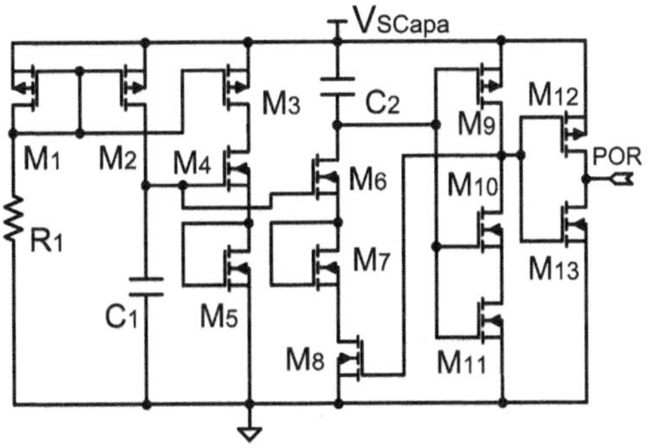

Fig. 2.48 Power-On Reset (POR) architecture

Table 2.13 Effect of the POR circuit in the power consumption reduction during the Start-up of the complex module

$V_{SCapa} = 0.6$ V	POR	No POR
P_{CC}	1 μW	6 μW

Furthermore, extra circuits can be added easily using some of the integrated circuits of the module. For instance, a temperature sensor is designed using the on-chip BG circuit in [93] where a voltage proportional to the temperature is delivered to the load.

The power consumption of the module is presented in Fig. 2.49, where it is compared with the typical simulated values. In that case the simulated and experimental results are really close. Finally, the layout of the Complex Module is presented in Fig. 2.50.

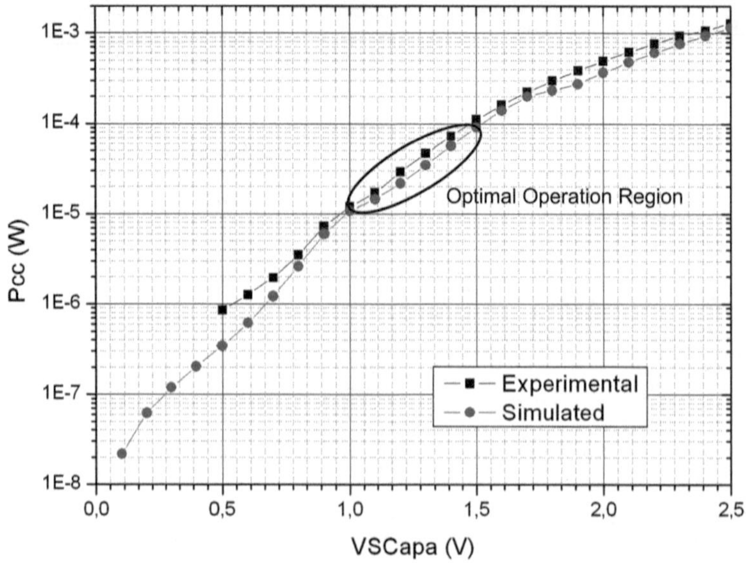

Fig. 2.49 Complex module power consumption

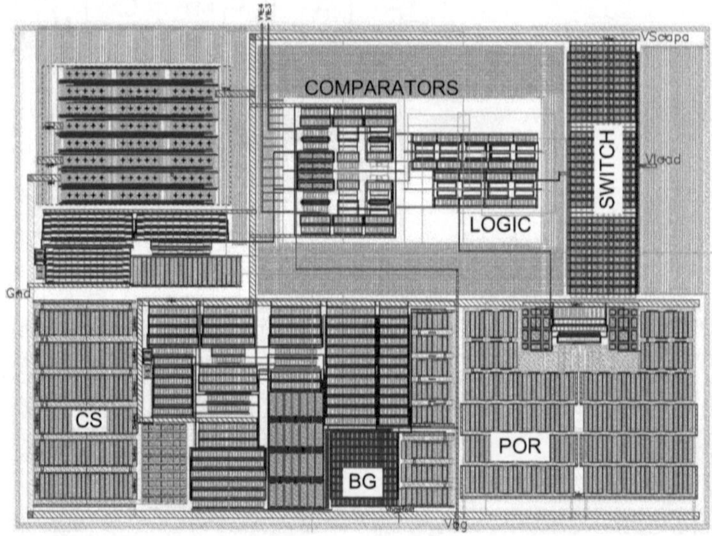

Fig. 2.50 Complex control module layout (770 μm × 545 μm)

Simple Control Module

The second control module is only based in a comparator, Fig. 2.51. The input branches and the decision stage define a high hysteresis cycle used to define the max and min voltage values. The same figure also presents the simulated hysteresis of 200 mV centered in 1.3 V defining a V_{min} of 1.2 V and V_{max} of 1.5 V.

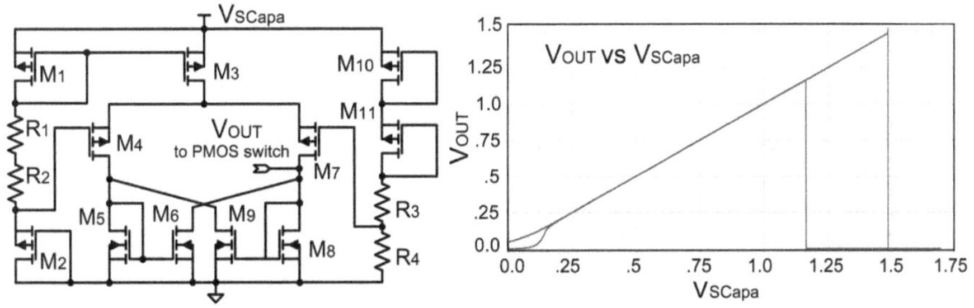

Fig. 2.51 Comparator architecture used in the Simple Module and its simulated hysteresis used to define the max and min threshold voltages

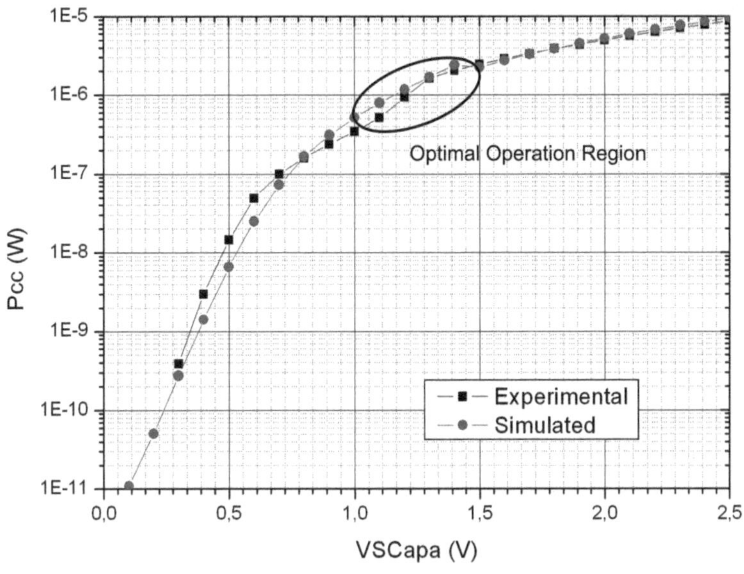

Fig. 2.52 Simple module power consumption

Its main advantage resides in the low power consumption and its small area (260 μm × 490 μm). Figure 2.52 presents the power consumption in function of several voltage values. As can be observed the consumption is 10 times lower than the one consumed by the Complex Module. On the other hand, a good layout matching is necessary to achieve a good precision and no programmability neither other extra features is included. Figure 2.53 depicts the experimental hysteresis and the final layout.

The typical drop voltage across the PMOS switch is 11 mV for both modules. Several Montecarlo simulations show that the drop range can vary from 25 mV to 3 mV.

Table 2.14 summarizes and compares the functions of each control module. In those applications where the power consumption is a must, the simple control is

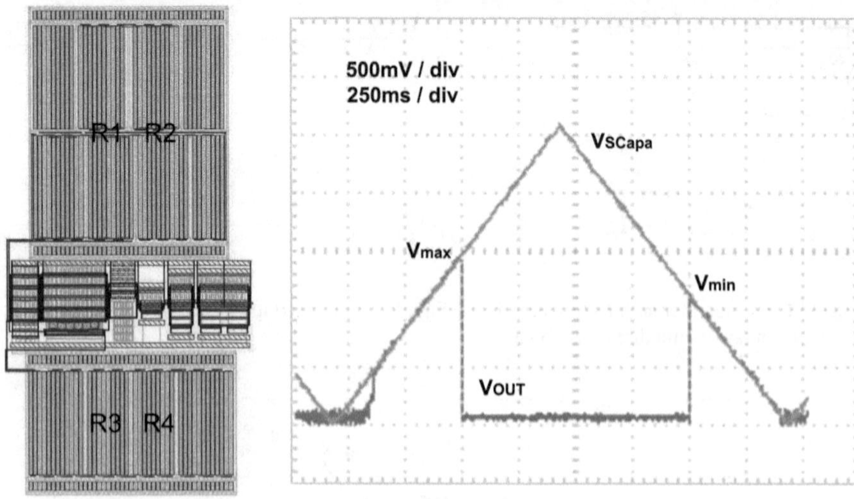

Fig. 2.53 Layout and experimental output voltage of the simple comparator

Table 2.14 Summary of the main characteristics for both control modules

		Modules				
		Complex		Simple		
Symbol	Parameter	Min	Max	Min	Max	Units
V_{SCapa}	Input voltage	0.9	2.5	0.9	2.5	V
P_{cc}	Power consumption	10 µ	1 m	0.5 µ	10 µ	W
V_{max}	Maximum threshold voltage	Programmable		Fixed		V
		1	2.5	1.5		
V_{min}	Minimum threshold voltage	Programmable		Fixed		V
		1	2.5	1.2		
Precision		High		Low		
Area		Large		Small		
Extra functionalities		Yes		No		

clearly the option to be selected; on the other hand, the complex module is the option to develop complex applications.

2.3.3.3 SPRCM Results

Both modules have been tested with a real piezoelectric generator and the experimental values have been compared with the simulated ones, anyhow only the results obtained with the complex module are presented below. The AC/DC NMOS rectifier (Section 2.3.2.2) is the selected option to rectify the AC signal provided by the piezo.

Fig. 2.54 Experimental setup

The experimental setup, Fig. 2.54, and the simulated generator are based on a Quick Pack piezoelectric QP40W from Midé Technologies working at 7 m/s2 @ 80 Hz. The simulated behavior of the piezo is based on a Lumped Electrical Model compatible with Spectre software explained in Section 2.3.1 and developed in detail in [17, 56].

The QP40W transducer is mounted over an electromagnetic vibration shaker ET-132 Labworks® Inc. The shaker is controlled with the Pa – 119 Labworks® Inc. amplifier. A LC2442 laser by Keyence® is used for the displacement measurements and two Tektronix® oscilloscopes TDS 2014 and TDS 714L are used to visualize the circuit waveforms.

An estimated load power consumption of 1.5 mW has been assumed to carry out with the simulations and experimental tests. It is also assumed that the power is consumed in pulses of 10 ms. The selected value of the storage capacitor is 47 μ F. A super capacitor or larger capacitance values are not used to avoid long simulation times.

Using the previous values and taking into account that the energy transferred from the capacitor to the load is described by (Eq. 2.13) where V_{max} and V_{min} are the defined max and min voltage values at the capacitor and C_{SCapa}, is the capacitance value of the storage device, the total energy involved in the Discharge Phase is around 5 μ J defining an average output current of 600 μ A. For this study 1.1 V and 1.0 V are used as a max and min voltage respectively. These values are obtained using the on-chip resistance network in the complex control module.

$$\Delta E_{(J)} = 0.5 C_{SCapa} (V_{\max})^2 - 0.5 C_{SCapa} (V_{\min})^2 \tag{2.13}$$

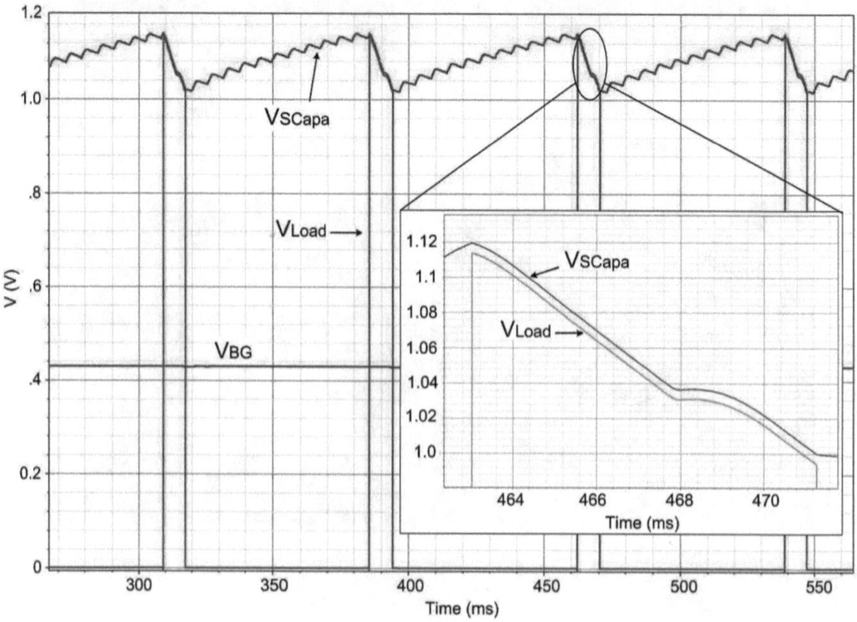

Fig. 2.55 Simulated steady state response and switch voltage drop (inset picture)

Several interesting values can be extracted from the simulations. The simulated Start-Up time is 230 ms and it is depicted in Fig. 2.43. The time involved in the Discharge Phase is 8 ms and in the Recharge phase is 60 ms (Fig. 2.55). All these values are for a typical (TYP) case. Furthermore, Fig. 2.55 depicts in detail the drop voltage of 11 mV across the switch.

Figure 2.56 presents the experimental Start-up. The time involved in this phase is 250 ms. The max voltage at the capacitor is 1.14 V whereas V_{min} is 1.02 V. The Bandgap voltage is 488 mV. Figure 2.57 presents the experimental Discharge and Recharge time of 10 ms and 55 ms respectively. The inset chart presents a zoom of the drop voltage of 17 mV across the PMOS switch.

The total consumption of the electronics is 67 μW. This value corresponds with those presented in Fig. 2.49. All experimental values are closed to the extracted simulated values and the system is fully validated. Moreover, more successful tests have been carried out using external resistors networks working with several values of V_{max} and V_{min} and using several capacitor values from 1 nF to 4,700 μF.

The maximum piezoelectric mechanical to electrical efficiency is obtained when $V_{CO} = V_{OC}/2$ [94] where V_{OC} is the open circuit voltage generated by the piezo and V_{CO} is the voltage generated by the PZT with load connected to them. Transduction efficiency around 75–80% is achieved with the system and it is independent of the capacitor value used as a SCapa thanks to the threshold programmability function of the complex control unit. It is easy to program the max and min voltage to obtain a voltage at the capacitor close to $(V_{max} + V_{min}) / 2 = V_{CO} = V_{OC} / 2$.

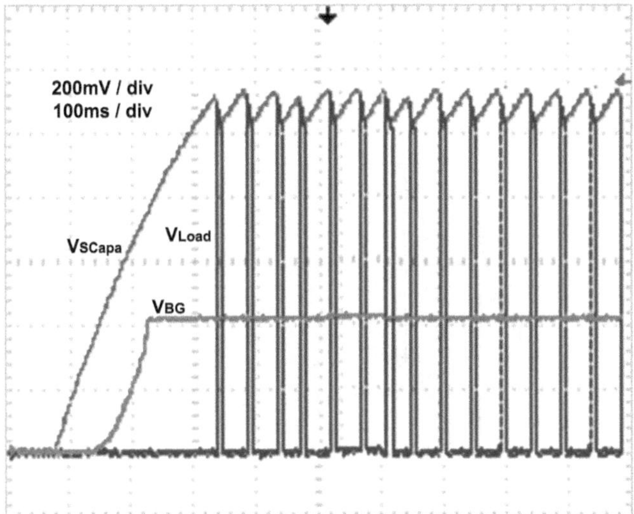

Fig. 2.56 Experimental complex module start-up

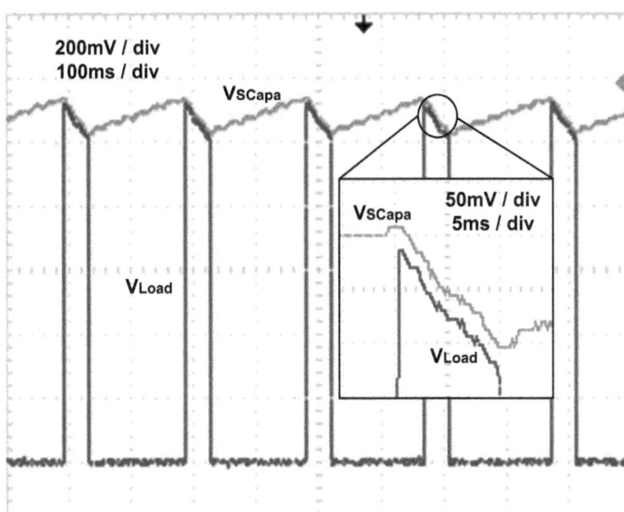

Fig. 2.57 Experimental steady state response and switch drop of the complex module

It is possible to observe in Fig. 2.58 the relation between V_{OC} and V_{CO}. It is also possible to observe the discharge / recharge phase of the Scapa (V_{SCapa}) and the voltage transferred to the load (V_{Load}). In that case the mechanical to electrical efficiency (η_{ME}) is around 75%, and the power efficiency of the module is around 80%.

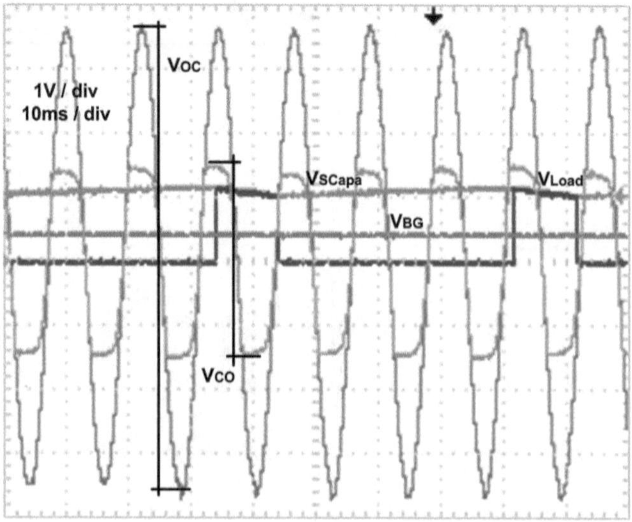

Fig. 2.58 Open circuit voltage (V_{oc}) and close circuit voltage (V_{co}) of the PZT working with the complex module

It is important to remark that similar waveforms are obtained for the simple control module. A reduction is observed in the Start-Up and recharge times due to its lower consumption. Moreover, there is the possibility to connect a voltage regulator like an LDO after the modules. In that way several modules can be connected in parallel working with the same storage element and different piezoelectric generators vibrating at different frequencies and places. It is stated the LDO or LDO's must work within the range defined by the max and min values of the modules.

2.4 Chapter Conclusions

This chapter discusses the design of a full custom Multi Harvesting Power Chip (MHCP) or, what is the same, a chip able to transform into electrical energy the power coming from three different ambient sources:

- mechanical vibrations,
- solar radiation and
- magnetically induced energy.

The MHPC is explained and detailed at the beginning of the chapter. The two modes of operation are introduced, the SSD (Single Storage Device), where only one storage device is shared for all the sources, and the MSD (Multiple Storage Device), where each source uses one storage device. Then, the ability of the circuit to work with all the sources working at the same time is validated and demonstrated.

The architecture of the power management circuit has been proposed and each circuit block described. Basically, the following circuits have been designed with a low-power low-voltage commercial technology (0.13 μ m STMicroelectronics):

- Low-DropOut (LDO) regulator and a Bandgap circuit.
- Full and half wave AC/DC Rectifier.
- Low-Voltage Charge Pump (CP).
- Two novel Self-Powered Regulation Control Modules (SPRCM).

Each circuit has been explained and experimentally validated in detail with special attention in the stability of the LDO and in the full MOSFET Bandgap voltage reference. Moreover, the novel SPRCM modules are exhaustively detailed and validated through simulations and experimental results with low-cost commercial piezoelectric generators.

High mechanical to electrical conversion efficiency are obtained with the SPRCM modules when working with piezoelectric generators. They are working continuously looking for the correct relation between open circuit and closed circuit voltages of the piezos.

It has also been introduced the modeling of the different sources. In that way, an inductive link formed by two coupled antennas and a solar cell t has been presented. Moreover, a simple real lumped piezoelectric model has been introduced, referenced and validated through experimental results.

References

1. S.J. Miller-Smith, New Chip Can Read Your Pet's Temperature, Darwin Veterinary Center. www.darwinvets.plus.com/topical/biothermo.htm
2. E.O. Torres, G.A. Rincón-Mora, Electrostatic energy-harvesting and battery-charging CMOS system prototype. IEEE J. Circuits Syst.–I, **56**(9), (Sept 2009)
3. S. Alepuz, S. Busquets-Monge, J. Bordonau, J. Gago, D. Gonzalez, J. Balcells, Interfacing renewable energy sources to the utility grid using a three-level inverter. IEEE Trans. Ind. Electron. **53**(5), 1504–1511 (Oct 2006)
4. J.M. Carrasco, L.G. Franquelo, J.T. Bialasiewicz, E. Galvan, R.C. PortilloGuisado, M.A.M. Prats, J.I. Leon, N. Moreno-Alfonso, Power-electronic systems for the grid integration of renewable energy sources: A survey. IEEE Trans. Ind. Electron. **53**(4), 1002–1016 (June 2006)
5. J. Schonberger, R. Duke, S.D. Round, DC-bus signaling: A distributed control strategy for a hybrid renewable nanogrid. IEEE Trans. Ind. Electron. **53**(5), 1453–1460 (Oct 2006)
6. L. Collins, Harvest for the world. IEE Power Eng. **20**(1), 34–37 (Feb–March 2006)
7. J.A. Paradiso, T. Starner, Energy scavenging for mobile and wireless electronics. IEEE Pervasive Comput. **4**(1), 18–27 (Jan–March 2005)
8. E.M. Yeatman. in. Energy scavenging for wireless sensor nodes. Proceedings of the 2nd International Workshop on Advances in Sensors and Interface, 2007. IWASI 2007, (2007) p. 4
9. S. Roundy, D. Steingart, L. Frechette, P. Wright, J. Rabaey, Power sources for wireless sensors networks. 1st European Workshop on Wireless Sensors Networks, (Berlin, Germany, 2004)

10. D. Niyato, E. Hossain, M.M. Rashid, V.K. Bhargava, Wireless sensor networks with energy harvesting technologies: A game-theoretic approach to optimal energy management. IEEE Wirel. Commun. **14**(4), 90–96 (Aug 2007)
11. T. Starner, J.A. Paradiso, in *Human-Generated Power for Mobile Electronics*, ed. by C. Piguet. Low-Power Electronics Design, (CRC Press, Boca Raton, 2004), Chapter 45, pp. 1–35
12. E.K. Reilly, E. Carleton, P.K. Wright, Thin film piezoelectric energy scavenging systems for long term medical monitoring, IEEE Proceedings of the International Workshop on Wearable and Implantable Body Sensor Networks, (2006), p. 4
13. X. Cao, W. Chiang, Y. King, Y. Lee, IC energy harvesting circuit with feedforward and feedback DC–DC PWM boost converter for vibration power generator system. IEEE Trans. Power Electron. **22**(2), 679–685 (March 2007)
14. http://www.ixys.com
15. http://www.mide.com
16. J. Brufau, M. Puig, Piezoelectric polymer model validation applied to mm size micro-robot I-SWARM (intelligent swarm), Proceedings of the SPIE 2006. 6166, pp. 229–240
17. L. Mateu, F. Moll, System –level simulation of a self-powered sensor with piezoelectric energy harvesting. Sensor technologies and applications, International conference on sensorComm, (2007) pp. 399–404
18. D. Puccinelli, M. Haenggi, Wireless sensor networks: applications and challenges of ubiquitous sensing. IEEE Circuits Syst. Mag., **3**(3), 19–29 (2005)
19. M. Ferrari, V. Ferrari, D. Marioli, A. Taroni, Modeling, fabrication and performance measurements of a piezoelectric energy converter for power harvesting in autonomous microsystems, IEEE Trans. Instrum. Meas. **55**(6), pp. 2096–2101, (Dec 2006)
20. http://www.ti.com
21. K. Van Schuylengergh, R. Puers, *Inductive Powering. Basic Theory and Application to Biomedical Systems*, (Springer, 2009), ISBN: 978-90-481-2411-4
22. B. Lenaerts, R. Puers, *Omnidirectional Inductive Powering for Biomedical Implants*, (Springer, 2009), ISBN: 978- 1- 4020-9074-5
23. M.Veefkind, Industrial design and PV-power, challenges and barriers, *Proceedings of the ISES EuroSun Congres*, (2004)
24. A. Reinders, Options for photovoltaic solar energy systems in portable products, *Proceedings of the 4th TCME*, (2002)
25. A. Bertacchini, D. Dondi, L. Larcher, P. Pavan, Perfomance analysis of solar energy harvesting circuits for autonomous sensors, *Proceedings of the 34th Annual Conference of IEEE Industrial Electronics*, 2008 (IECON), (2008) ISBN: 978-1-4244-1767-4
26. S.K. Lau, K.N. Leung, P. Mok, Analysis of Low-Dropout Regulator Topologies for Low-Voltage Regulation, IEEE Conference on Electron Devices and Solid-State Circuits, pp. 379 – 382, 2003.LDO1
27. B.M. King, Advantages of using PMOS-type low-dropout linear regulator in battery applications. Analog Appl. J., Texas Instruments, Inc. (2000) Available at: focus.ti.com/lit/an/slyt161/slyt161.pdf
28. J.A. Cueto, T.J. McMahon, Performance of single-junction a-Si modules under varying conditions in the field, 26th IEEE Photovoltaic specialists conference, (1997)
29. M. Green, Photovoltaic: Technology overview. Energy Policy. **28**, 989–998 (2000)
30. C. Sauer, M. Stanacevic, G. Cauwenberghs, N. Thakor, Power harvesting and telemetry in CMOS for implant devices. IEEE Trans. Circuits Syst. **52**, 12 (Dec 2005)
31. N. Utsuyama, H. Yamaguchi, S. Obara, H. Tanaka, S. Fukutaand, J. Nakahira, S. Tanabe, E. Bando, H. Miyamoto, Telemetry of human electrocardiograms in aerial and aquatic environments, IEEE Trans. Biomed. Eng. **35**, 881–884 (1988)
32. J. Van Ham, P. Reynders, R. Puers, *An autonomous implantable distraction nail controlled by an inductive power and data link*, Transducers 2007 and Eurosensors XXI, (Lyon, France, 2007). pp. 427–430.

33. Zarlink Semiconductors Inc, ZL70101Medical Implantable RF Transceiver, (2007)
34. U.S. Inan, A.S. Inan, *Electromagnetic Waves: Electromagnetic Radiation and Elementary Antennas*, (Prentice Hall, 2000), pp. 476–499, ISBN 0-201-36179-5
35. L.W. Hart, H.W. Ko, J.H. Meyer, D.P. Vasholz, R.I. Joseph, A noninvasively electromagnetic conductivity sensor for biomedical applications. IEEE Trans. Biomed. Eng. **35**, 1011–1022 (1988)
36. E.S. Hochmair, System optimization for improved accuracy in transcutaneous signal and power transmission. IEEE Trans. Biomed. Eng. **31**, 177–186, (1984)
37. F.S. Barnes, *Interaction of DC and ELF electric fields with biological materials and systems*, Handbook of Biological Effects of Electromagnetic Fields, (CRC Press, 1996), 2nd edn., chapter. 2, pp. 103–148, ISBN: 0-8493-0641-8
38. Microchip Application Note, Antenna Circuit Design for RFID Applications AN710, (2003)
39. ST Microelectronics Application Note, Antenna (and Associated Components) Matching – Circuit Calculation for the CRX14 Coupler, AN1806, (Feb 2006)
40. S. Serkan, K. Lim, J. Laskar, M. Tentzeris, Design and modelling of embedded 13.56 M Hz RFID antennas., IEEE – Antennas and Propagation Society International Symposium, **4B**, 64 – 67 (July 2005), ISBN: 0-7803-8883-6
41. Texas Instruments datasheet, Stability analysis of Low-dropout linear regulators with PMOS pass element, SLYT194, 1999
42. R. Tantawy, E.J. Brauer. Performance Evaluation of CMOS Low Drop-Out Voltage Regulators, Proceedings of the 47th IEEE International Midwest Symposium on Circuits and Systems, (2004)
43. K. Nang Leung et al. A Low-voltage CMOS low-dropout regulator with enhanced loop response, Proceedings of IEEE ISCAS 2004 pp. 385–388
44. R. Jacob Baker, *CMOS: Circuit Design, Layout, and Simulation. Revised second edition*, (Wiley – Interscience, 2008), ISBN 978–0-470-22941-5
45. R. Tantawy, E.J. Brauer, Performances Evaluation of CMOS Low Brop-Out Voltage Regulators, Proceedings of the 47th IEEE International Midwest Symposium on Circuits and Systems (MWSCAS), (2004)
46. Giustolisi, et al. A low-voltage low-power voltage reference based on subthreshold MOSFET's. IEEE J. Solid-State Circuits. **38**(1), 151–154 (2003)
47. M.D. Ker, J.S. Chen. New Curvature-Compensation Technique for CMOS Bandgap Reference with Sub-1-V Operation. IEEE Trans. Circuits and Syst. II. **53**(8), 667–671 (Aug 2006)
48. K. Sanborn, M. Dongsheng, V. Ivanov, A sub-1-V low-noise bandgap voltage reference. IEEE J Solid-State Circuits. **42**(11), 2466–2481 (Nov 2007)
49. D.V. Kerns, Optimization of the peaking current source. IEEE J. Solid-State Circuits. **21**(2), 587–590 (1986)
50. D.V. Kerns, Enhanced peaking current reference. IEEE J. Solid-State Circuits. **23**(3), 869–872 (1988)
51. S. Chunlei, B.C. Walker, E. Zeisel, B. Hu, G.H. McAllister. A Highly Integrated Power Management IC for Advanced Mobile Applications. IEEE Trans. Solid-State Circuits. **42**(8), 1723–1731 (Aug.2007)
52. M.H. Cheng, Z.W. Wu. Low-power low-voltage reference using peaking current mirror circuit. Electron. Lett. **41**(10) (May 2005)
53. M.H. Rashid, *Power Electronics Handbook*, (Academic Press), ISBN 0120884798
54. M.H. Rashid, *Power Electronics: Circuits, Devices and Applications* 3rd edn., (Academic Press), ISBN-10: 0125816502
55. F. Kocer, P.M. Walsh, M.P. Flynn, Wireless, remotely powered telemetry in 0.25 μm CMOS, radio frequency integrated circuits symposium, (IEEE Press, 2004), pp. 339–342
56. J. Brufau, *Smart materials for microrobotics. Motion Control and Power Harvesting*, (Universitat de Barcelona, Barcelona, 2009)

57. J. Colomer-Farrarons, P. Miribel-Català, A. Saiz-Vela, M. Puig-Vidal, J. Samitier, Power-conditioning circuitry for s self-powered system based on micro pzt generators in a 0.13 μ m low-voltage low-power technology. IEEE Trans. Ind. Electron. **55**(9), (Sept 2008), ISSN 0278-0054
58. L. Mateu, *Energy Harvesting from Human Passive Power*, (Universitat Politecnica de Catalunya, Catalonia, 2009)
59. J. Brufau, M. Puig, Piezoelectric energy harvesting improvement with complex conjugate impedance matching. J. Intell. Mater. Syst. Struct. 2009, 597–608 (Sep 2008), DOI: 10.1177/1045389X08096051
60. G. Patounakis, Y.W. Li, K.L. Shepard, A fully integrated on-chip DC-DC conversion and power management system. IEEE J. Solid-State Circuits. **39**, 3 (Jan 2004)
61. A.M. Abo, P.R. Gray, A 1.5-V, 10-bit, 14.3-MS/s CMOS pipeline analog-to-digital converter. IEEE J. Solid-State Circuits. **34**(5), (May 1999)
62. D. Aksin et al., A bootstrapped switch for precise sampling of inputs with signal range beyond supply voltage, IEEE Custom Integrated Circuits Conference, pp. 743–746 (2005)
63. ST Microelectronics: http://www.st.com
64. Europractice service: http://www.europractice.com/
65. CMP service: http://cmp.imag.fr/
66. R. Gregorian, *Introduction to CMOS Op-Amps and Comparators* (Wiley & Sons, USA, 1999), ISBN 0-471-31778-0
67. R.J. Milliken, *A Capacitor – Less Low Drop – Out Voltage Regulator with Fast Transient Response* (Texas A&M University, PhD Thesis, December 2005) Available at: http://repository.tamu.edu/handle/1969.1/3275
68. P. Corbishley, E. Rodriguez – Villegas, A low power low voltage rectifier circuit. IEEE Int. Midwest Symp.Circuits Syst. MWSCAS'06". **2**, 512–515 (2006), ISBN: 1-4244-0172-0
69. A.J. Cardoso, C.R. Rodrigues, R.S. Pippi, A.Cesar, F.C.B. Vieira, CMOS Energy Harvester Based on a Low-Cost Piezoelectric Acoustic Transducer, Proceedings of the 49th IEEE International Midwest Symposium on Circuits and Systems, 2006. MWSCAS ′06. Vol 1, (6–9 Aug. 2006) pp. 70–74
70. R. Fiorelli, A. Arnaud, C. Galup-Motoro, Series-Parallel Association of Transisotrs for the Reduction of Random Offset in Non-Unity Gain Current Mirrors, Proceedings of the International Conference on Circuits and Systems ISCAS'04, Vol. 1 (May 2004), ISBN: 0-7803-8251-X
71. M. Ferrari, V. Ferrari, M. Giuzzetti, D. Marioli, An autonomous battery-less sensor module powered by piezoelectric energy harvesting with RF transmission of multiple measurement signals, Smart Materials and Structures
72. T. Lehman, Y. Moghe, On–Chip active power rectifiers for biomedical applications. IEEE Int. Symp. Circuits Syst. 2005 ISCAS'05. **1**, 732–735 (2005), ISBN: 0-7803-8834-8
73. M. Huang et. al., Sub-1 V Input Single-Inductor Dual-Output (SIDO) DC-DC Converter with Adaptive Load-Tracking Control (ALTC) for Single-Cell-Powered System, Proceedings of the European Solid-State Circuit Conference (ESSCIRC'09), (2009) pp. 268–271
74. P.B. Zbar, Prácticas de electrónica, Marcombo S. A., (1988), ISBN 84-267-0233-6
75. K. Bhattacharyya, P.V. Ratna Kumar, P. Mandal, Embedded Hybrid DC-DC Converter with Improved Power Efficiency, Proceedingsof the 52nd International Midwest Symposium on Circuits and Systems, (2009) pp. 945–948
76. A. Saiz-Vela, P. Miribel-Català, J. Colomer-Farrarons, J. Samitier, Ripple reduction on skipping-based regulated two-phase voltage doubler charge pump. Electron. Lett. **45**, 20 (Sept 2009)

77. R. Pelliconi, D. Iezzi, A. Baroni, M. Pasotti, P. Rolandi, Power efficient charge pump in deep submicron standard CMOS technology. IEEE J. Solid-State Circuits. **38**(6), 1068–1071 (June 2003)
78. J. Soldera, A. V. Boas, A. Olmos, A Low Ripple Fully Integrated Charge Pump Regulator, Proceedings of the 16th Symposium on Integrated Circuits and System design, (2003) pp. 170–180
79. A. Saiz-Vela, P. Miribel-Català, J. Colomer-Farrarons, M. Puig-Vidal, J. Samitier, Accurate design of high multistage voltage doublers based on compact mathematical model. Electron. Lett. **43**, 15 (July 2007)
80. J.A. Starzyk, J. Yin-Wei, Q. Fengjing, A DC-DC Charge pump design based on voltage doublers. IEEE Trans. Circuits Syst.I Fundam. Theory Appl. **48**(3), 350–358 (March 2001)
81. C. Chan; W.H. Ki; C.Tsui; Bi-directional integrated charge pumps, IEEE International Symposium on Circuits and Systems, (2002), 827–830, ISBN: 0-7803-7448-7
82. E. Bayer, H. Schemeller, Charge Pump with Active Cycle Regulation-Closing the Gap Between Linear and Skip Modes, Proceedings of the IEEE International Power Electronics Specialists Conference, (PESC), (June 2000), pp. 1497–1502
83. J. Doutreloigne, H. de Smet, J. Van den Steen, G. Van Doorsealer, Low-Power High-Voltage CMOS Level-Shifters for Liquid Crystal Display Drivers, Proceedings of the 11th International Conference on Microelectronics, (1999), pp. 13–216
84. A. Saiz-Vela, P. miribel-Català, J. Colomer-Farrarons, M. Puig-Vidal, J. Samitier, AccuRate Design of Two-Phase Voltage Doublers Based on a Compact Mathematical Model, Proceedings of the 50th Midwest Symposium on Circuits and Systems (MWSCAS'07), (2007), pp. 213–216
85. I. Doms, P. Merken, C. Van Hoof, R. P. Mertens, Capacitive power management circuit for micropower thermoelectric generators with a 1.4 μA controller. IEEE J. Solid State Circuits. **44**, 10 (Oct 2009)
86. F. Kocer, P.M. Walsh, M. P. Flynn, *Wireless Remotely Power Telemetry in 0.25 μm CMOS*, 2004 VLSI Symposium, (June 2004) pp. 24–27
87. T.S. Salter Jr., G. Metze, N. Goldsman, Improved RF Power Harvesting Circuit Design, International Semiconductor Device Research Symposium, (2007)
88. W. Leran, T.J. Kazmierski, B.M. Al-Hashimi, S. P. Beeby, R. N. Torah, Integrated approach to energy harvester mixed technology modeling and performance optimization, Design, Automation and Test in Europe (DATE'08), (2008)
89. http://www.peltron.com
90. http://www.cymbet.com
91. http://www.silicore.com
92. W. C. Yen, H. W. Chen, Low power and fast system wakeup circuit. IEE Proceedings on Circuits, Devices and Systems, **152**, 223–228 (2005)
93. J. Colomer-Farrarons et. al., Self-Powered Temperature Sensor Powered by Vibration Energy Harvesting, Proceedings of the 32nd Conference on Design of Circuits and Integrated Systems, 2007, ISBN-13 978-84690-8629-2
94. E. Lefeuvre et al., Piezoelectric energy harvesting device optimization by synchronous electric charge extraction. J. Intell. Mater. Syst. Struct. **16**, 865–876 (2005)
95. N.S. Shenck and J.A. Paradiso, Energy scavenging with shoe-mounted piezoelectrics. IEEE Micro. **21**(3), 30–42 (2001)
96. M.S.M. Soliman, E.F. El-Saadany, R.R. Manssur. Electromagnetic MEMS based micro-power enerator. 2006 IEEE International Symposium on Industrial Electronics. **4**, 2747–2753 (July 2006)
97. S. Meninger, J.O. Mur-Miranda, R. Amirtharajah, A. Chandrakasan, J.H. Lang Vibration-to-electric energy conversion. IEEE Trans. Very Large Scale Integration (VLSI) Syst. **9**(1), 64–76 (2001)

98. N. Amor, O. Kanoun, Investigation to the Use of Vibration Energy for Supply of Hearing Aids, Proceedings of the 2007 IEEE Instrumentation and Measurement Technology Conference,(1–3 May 2007), pp. 1–6
99. N. Cho et al., A 8-μW, 0.3 mm2 RF-Powered Transponder with Temperature Sensor for Wireless Environmental Monitoring, Proceedings of the Int'l Symposium Circuits and Systems (ISCAS 2005), (IEEE Press, 2005), pp. 4763–4766
100. N. Hayakawa, A study of the new energy system for quartz watches (II)-The effective circuit for the system, in Proceedings of the Congrès Eur. Chronomét., (1988), pp. 81–85
101. J. Yoshida, Piezoelectric rackets add professional oomph, in Electronic Engineering Times. (Manhasset, NY: CMP), (Jun 10, 2002)
102. C.B. Williams, R.B. Yates, Analysis of a micro-electric generator for microsystems. Sens. Actuators A **52**, 8–11 (1996)
103. M. El-hami, P. Glynne-Jones, N.M. White, M. Hill, S. Beeby, E. James, A.D. Brown, J.N. Ross, Design and fabrication of a new vibration-based electromechanical power generator. Sens. Actuators A **92**, 335–342 (2001)
104. T. Starner, Human-powered wearable computing. IBM Syst. J. **35**, 618–629, (1996)
105. J. Colomer-Farrarons, J. Brufau, P. Miribel-Catala, A. Saiz-Vela, M. Puig-Vidal, J. Samitier Power Conditioning Circuitry for a Self-Powered Mobile System Based on an Array of Micro PZT Generators in a 0.13A0~μm Technology, Proceedings of the IEEE International Symposium on Industrial Electronics, 2007. ISIE 2007. (4–7 June 2007) pp. 2353–2357
106. M. Ghovanloo, K. Najafi, *Fully integrated power supply design for wireless biomedical implants*, 2nd Annual International IEEE – EMB Conference on Microtechnologies in Medicine & Biology, (IEEE Press, 2002), pp. 414–419, ISBN 0-7803-7480-0
107. Q.T. Hoang, B.F. Hete, R.P. Gaumond, Comments on radio-frequency coil in implantable devices: misalignment analysis and design procedure. IEEE Trans. Biomed. Eng. **35**, 1011–1022 (1988)
108. Texas Instruments datasheet, A low-voltage CMOS low-dropout Regulator, LP2985, (June 2001)
109. E. Dalago, D. Miatton, G. Venchi, V. Bottarel, G. Frattini, G. Ricotti, M. Schipani, Electronic Interface for Piezoelectric energy Scavenging System, Proceedings of the 34th European Solid – State Conference ESSCIRC 2008, pp. 402 – 405, ISBN: 978-1-4244-2361-3
110. Y. Ammar et al., Wireless sensor network node with asynchronous architecture and vibration harvesting micro power generator, *Proceedings of the 2005 joint conference on Smart objects and ambient intelligence*, pp. 287–292, ISBN: 1-59593-304-2
111. T.S. Salter Jr., G. Metze, N. Goldsman, Improved RF power harvesting circuit design. ISDRS'07 Conf. (Dec 2007)
112. L. Wang, et al., *Integrated approach to energy harvester mixed technology modelling and performance optimisation*, University of Southampton, ISBN: 978-3-9810801-3-1
113. A. Facen, A. Boni, *Power Supply Generation in CMOS Passive UHF RFID Tags*, Research in Microelectronics and Electronics (2006), pp. 33–36, ISBN: 1-4244-0157-7
114. C. Sauer, M. Stanacevic, G. Cauwenberghs, N. Thakor, Power Harvesting and telemetry in CMOS for implanted devices. IEEE Trans. Circuits and Systems-I. **52**(12), (Dec 2005)
115. P. Vaillancourt, A. Djemouai, J.F. Harvey, M. Sawan, EM radiation behaviour upon biological tissues in a radio-frequency power transfer link for a cortical visual implant. Proc. IEEE EMBS Conf. **6**, 2499–2502 (1997)
116. K. Myny et al. An inductively – coupled 64b organic RFID tag operating at 13.56 M Hz with a data rate of 787b/s, IEEE International Solid State Circuits Conference, pp. 290–291 (2008)
117. M. Barú, O. de Oliveira, F. Silveira, A 2 V Rail – to – Rail Micropower CMOS Comparator, Proceedings of the 11th Conference of the Brazilian Microelectronics Society, (Sep 1996)
118. Maxim-IC Application Note, Adding Extra Hysteresis to Comparators, AP 3616, (Aug 2005)

119. C. Huang, J. Wang, High-Performances and power–efficient CMOS comparators. IEEE J. Solid – State Circuits. **38** 2 (Feb. 2003)
120. T. Umeda et al., A 950-MHz rectifier circuit for sensor network tags with 10-m distance. IEEE J. Solid-State Circuits. **41**, 1 (Jan. 2006)
121. T.T. Le, J. Han, A. von Jouanne, K. Mayaram, T.S. Fiez, Piezoelectric micro-power generation interface circuits. IEEE J. Solid-State Circuits. **41**, 6 (June 2006)
122. E.O. Torres, G.A. Rincon-Mora, Energy-harvesting system-in-package (SiP) microsystem. ASCE J. Energy Eng. **134**(4), 121–129 (Dec 2008)
123. M. Marzencki, Y. Ammar, S. Basrour, Integrated power harvesting system including a MEMS generator and a power management circuit. Sens. Actuators A Phys. **145–146**, 363–370 (2008)
124. T. S. Salter Jr., G. Metze, N. Goldsman, *Improved RF Power Harvesting Circuit Design* ISDRS 2007, (College Park, MD, 2007)
125. L. Wang, T. Kazmierski, B. Al-Hashimi, S. Beeby, R. Torah, Integrated approach to energy harvester mixed technology modelling and performance optimization, Design, Automation and Test in Europe (DATE'o8), (2008)
126. R.P. Areny, Sensores y Acondicionadores de Señal, 3a edición, Marcombo S. A., (1998), ISBN 84-267-1171-5
127. D.W. Hart, *Electrónica de Potencia* (Prentice Hall, Spain, 2001), ISBN 84-205-3179-0
128. J. Colomer-Farrarons, P. Miribel-Català, A. Saiz-Vela, J. Samitier, A 60μW Low-Power Low-Voltage Power Management Unit for a Self-Powered System Based on Low-Cost Piezoelectric Powering Generators, Proceedings of the European Solid-State Circuit Conference (ESSCIRC'09), (2009) pp. 280–283

Chapter 3
Biomedical Integrated Instrumentation

Abstract This Chapter is focused on the development of an integrated instrumentation to work with three electrodes amperometric Biosensor. First of all, it is introduced the conception of three electrodes configuration and how it works. Moreover, some typical electrochemical techniques like Voltammetry, EIS and amperometry, are introduced to the reader. The instrumentation electronics is based on a potentiostat architecture, which is explained in detail and experimentally validated. The obtained results with the full-custom approach are compared with the ones obtained using a commercial potentiostat. In that way, the correct operation of the designed circuits is fully validated. Furthermore, this chapter explains the conception of a Lock-In amplifier circuit used to detect the real and imaginary components of the complex impedance measured from the Biosensor. This circuit is theoretically explained and some simulated results are shown. Finally, the conception of Biotelemetry or how to transmit information from the subcutaneous device to the external reader is introduced. Then, the implemented protocol in this work is detailed. In summary, this chapter presents the developed BioChip IC that is able to drive the sensor, process the measured data and transmit the data to the external side through an inductive link.

Keywords Analog integrated circuits · Microelectronic implants · Bioimpedance · Amperometric sensors · Electrochemical impedance spectroscopy · Phase detection · Active filters · Biomedical telemetry

3.1 General Introduction to Biomedical Instrumentation

A biomedical instrumentation system comprises [1], basically, the modules presented in Fig. 3.1. Its general architecture is similar to other conventional instrumentation system [2, 3]. The main difference between the bioinstrumentation and the conventional ones is the sort of signals to be analyzed. In Biomedical Systems (BS), the signals come from activities related with the human body and they have to be measured in-vivo through the living tissue, or using microelectrodes to detect biological activities [4, 5] or in-vitro, outside the human body [6].

J. Colomer-Farrarons, P.L. Miribel-Català, *A CMOS Self-Powered Front-End Architecture for Subcutaneous Event-Detector Devices*, DOI 10.1007/978-94-007-0686-6_3,

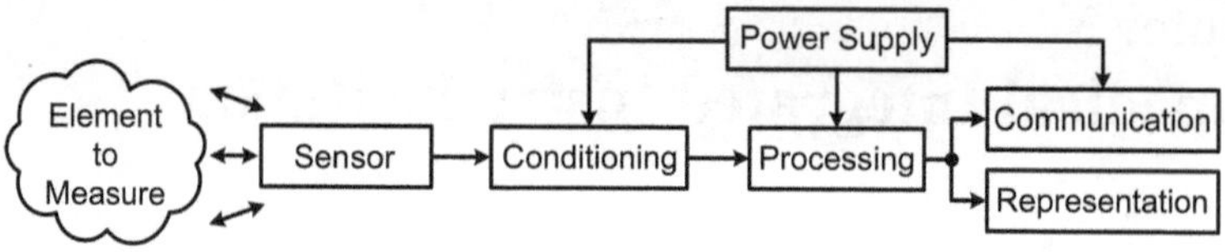

Fig. 3.1 General instrument architecture

The *Element to Measure or "Measurand"* [2, 3] is the value, property or physical quantity to be sensed and detected. In biomedical instrumentation this element could be a substance concentration, biopotential, pressure, impedance, etc...

The *Sensor* module is in charge to convert the energy generated by the *Measurand* to an electrical form [2, 4, 7, 8–10, 11]. This module is also known as transducer. There are several types of Biosensors, from the typical skin electrodes [5, 6, 12] used in electrocardiograms, till the miniaturized electrodes used to detect glucose, ADN ..., but all of them have to be specifically designed to extract the minimum amount of energy to be minimally invasive with their surroundings.

The output of the sensor is sometimes not enough valid to be processed. Hereby, the *Signal Conditioning* module prepares the output to be treated and analyzed. It could amplify, filter and even digitalize the incoming signal in order to prepare it for the following step. Once the signal is conditioned, the *Signal Processing* module evaluates the data and takes the necessary actions. At this stage, the data is organized and it could be displayed in a monitor or in the same device, or the module could activate an alarm if a threshold value is exceeded. Moreover, the *Signal Processing* module may prepare the data to be transmitted to other locations like a PC or a remote host.

Finally, if a communication is required, the *Communication* module is in charge to do that. It transmits the data prepared by the processing unit to the receiver or receivers using the established protocols, like TCP/IP, series, parallel, etc... It can transmit the information using cables or providing a wireless link between the medical instrument and the receiving unit, known as a Biotelemetry. It is a really useful tool since it allows transmitting information about one patient, i.e. in a hospital room, to the monitoring center without using the lengthy and cumbersome cables and giving full autonomy to the patient.

The energy in Fig. 3.1 is provided by the *Power Supply* unit. It manages the main energy source, battery, power line, harvesting, and delivers the necessary voltage and current to each module. It has to be carefully designed in order to guarantee the best isolation between the energy source and the electronic modules and *Measurand*, i.e., the patient. For implanted subcutaneous devices, inductive powering is one of the most useful methods to power the internal circuitry. The architecture shown above is not unique. It could be adapted by eliminating, adding or replacing some modules. Sometimes, *Processing* and *Communication* are joined in a single module and an extra calibration unit is added [13] in portable equipment. Moreover, some control feedback is included to adjust automatically drifts due to temperature or sensors degradation.

If the developed instrument is for implantable applications, inside human body, some extra considerations have to be done. First of all, the size of the instrument is very important. Given the human body dimensions, instruments not higher than a few cm are mandatory. Moreover, the consumption of those devices have to be really small hence, low-voltage low-power integrated instruments are the solution.

A perfect isolation between the instrument and its biological environment is a must in order to avoid biocompatibility and thermal problems. A temperature variation between one or two Celsius degree could produce injuries in the body and sensing problems.

3.2 Electrochemical Biosensors

A Biosensor is a type of sensor able to convert the biochemical interactions produced in a biological material into electrical or optical signals [14] allowing the detection of substances like enzymes [15, 16], antibodies [17, 18], proteins [19], etc...

Basically, it is formed by two parts: the Bioreceptor or biological detection element, which recognizes the element to be studied (enzymes, antibodies, etc...) and the Transducer that detects the biological reaction in the Bioreceptor and generates a proportional electrical signal. Depending on the type of Transducer the response of the Biosensor could be Potentiometric [19, 20], Amperometric [16, 21], Conductometric [22, 23], etc...

The electronics developed here are focused on working with low-frequency Amperometric [24, 25] Biosensor. An Amperometric sensor generates a current when a potential is applied on the electrochemical cell, which is proportional to the electrochemical reactions produced in the Bioreceptor. On the other hand, capacitive Biosensor produces a variation on its capacitance when a reaction takes place in the Bioreceptor hence, an impedance variation is produced in the electrode-solution interface when an AC signal is applied on the cell [14]. This means that an amperometric sensors delivers the electrochemical information by modulating the current through it, whereas, on Capacitive sensors the information is done by modification of the sensor impedance.

In general, amperometric and capacitive sensors are based on the use of 2 or 3 electrodes, Fig. 3.2. The most used topology is the three electrodes cell because it allows a better control of the applied potential to the cell. So, the developed instrumentation is for 3-electrodes sensors.

The function of each electrode is described below:

- *Counter* or *Auxiliar Electrode* (C/A): it supplies the necessary current required for the reaction at the W electrode.
- *Reference Electrode* (R): it is used to measure the potential between the *Reference* and *Working* electrodes in order to control the biochemical reaction.
- *Working Electrode* (W): it is covered with a sensitive film that serves as a surface where the reaction takes place.

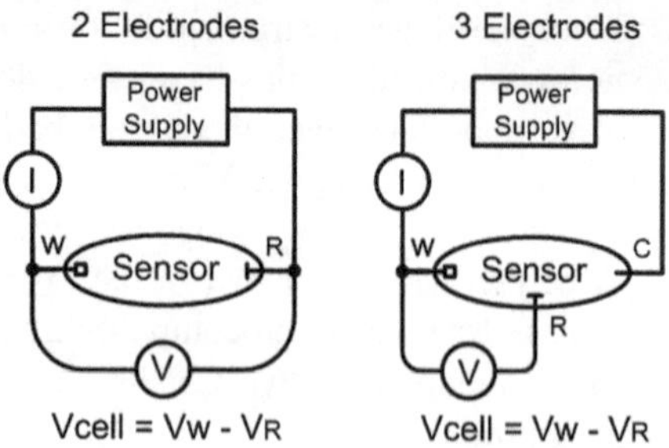

Fig. 3.2 Two and three electrodes sensor configuration

Mainly, the three electrodes sensor works as follows: a current flow (I, Fig. 3.2) is generated between C and W electrodes when a potential is applied between them (Power Supply). However, the electrochemical reaction starts on the W electrode when a well-known potential (V) is established between the *Reference* and *Working* electrodes.

A specific electronic circuit, known as potentiostat, is used to control the voltage between the R and W and to inject the current through the *Counter* electrode. Furthermore, it also measures the current flowing in the *Working* electrode. There are several options to develop this kind of circuits, from full analog solutions to mixed implementations. In that work, the integrated potentiostat is based on an analog development using operational amplifiers.

3.2.1 Three Electrodes Sensor Model

An electrochemical reaction could be described as a network of resistances, capacitances or impedances and active elements [26, 27]. These electrical equivalent circuits are important to understand the behavior and the working principle of the sensors. Moreover, these models are also used to design and test the sensor's electronics.

The representation of the interface model can be very complex [27]. The simplest case uses an equivalent circuit called Randles circuit [28]. It is presented in Fig. 3.3 and it is composed by the double-layer capacitor (C_{ref}), the polarization resistor (R_{ref}), which is also described as a charge transfer resistor, and the solution resistor (R_{AUX}).

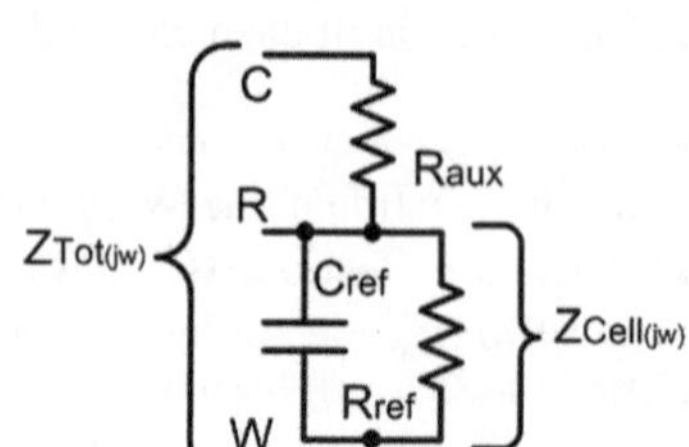

Fig. 3.3 Electrochemical Randles sensor model

These parameters are time-dependent and they have to be roughly known with an estimation that extends a couple of decades [29]. Then, the values of these parameters are used to guarantee stable operation of the implemented electronic circuit.

An important element is the presence of the capacitor, which is used to model the interface between the electrode and their surrounding electrolyte. This capacitive Biosensor is based on the theory of the electrical double-layer [30]. An electrode immersed in an electrolyte solution can be generally described as a capacitor in its ability to store charge. The double layer capacitor is formed as ions from a solution absorbed on the electrode surfaces.

The total impedance of the equivalent sensor model presented in Fig. 3.3 is defined by (3.1)

$$Z_{Tot}\,(jw) = \frac{V_{WC}(jw)}{I_C(jw)} = R_{aux} + \frac{R_{ref}}{1 + R_{ref}C_{ref}(jw)} \tag{3.1}$$

whereas the sensor Cell Impedance, where the reaction takes place, is defined by (3.2)

$$Z_{Cell}\,(jw) = Z_{W-R}\,(jw) = \frac{V_{WR}(jw)}{I_W(jw)} = \frac{R_{ref}}{1 + R_{ref}C_{ref}(jw)} \tag{3.2}$$

From (3.2), the real and imaginary parts of the impedance are derived.

$$Z_{Cell}\,(jw) = Z_{W-R}\,(jw) = \underbrace{\frac{R_{ref}}{1 + R_{ref}^2 C_{ref}^2 w^2}}_{Z_{REAL}} - \underbrace{\frac{R_{ref}^2 C_{ref} jw}{1 + R_{ref}^2 C_{ref}^2 w^2}}_{Z_{IMAG}} \tag{3.3}$$

Using this mathematical approach, it is possible to identify each electrical component of the model and relate each value with the corresponding electrochemical effect in the reaction. If there is some reaction in the sensor, automatically the corresponding electrical parameter will be modified and the impedance will change. Moreover, complex models can be developed just connecting, in series or parallel, more electrical components [31] to accurately describe the sensor behavior and some effects like Warburg resistance [32].

From Eqs. (3.1) and (3.3) can be deduced that applying an AC voltage signal (frequency dependent) to the sensor, it is possible to extract the real and imaginary part of the whole impedance.

On the other side, when a DC signal is applied just the real part is extracted. This means that it is possible to deduce all the impedance components combining DC and AC analysis techniques. These techniques, as well as their representations, are explained in the next section.

3.2.2 Cyclic Voltammetry, Amperometry, Impedance Analysis (EIS) and Their Representations

Several analysis techniques can be carried out with Biosensors in order to extract information about the analyzed substance. Some of the most used techniques are cyclic voltammetry [33, 34], amperometry [35], potentiometric analysis and EIS (Electrochemical Impedance Spectroscopy) [36]. Each technique will be applied in function of the type of sensor and substance to be analyzed.

The instrumentation developed here is designed to work with cyclic voltammetry and amperometric analysis and also prepared to perform low-frequencies *EIS* analysis.

The *Cyclic Voltammetry (CV)* method is widely used to study the products of electrochemical Reduction – Oxidation (Redox pair) reactions in the electrode/solution interface. It applies a triangular potential (Fig. 3.4a) between the R – W electrodes while the current through the *Working* is measured.

Then, this current is represented in function of the applied voltage (Fig. 3.4b). In that analysis, the max and min voltages, as well as the slope of the triangular wave, have influences in the reaction [37].

At the beginning of the analysis (point a, Fig. 3.4b) the generated current is small. Then, the current increases because the voltage approximates to the reduction potential (ab) and the electrons are forced to move from the electrode toward the analyzed substance. The maximum transfer is achieved at the Reduction potential (b).

Once the voltage overcomes the Reduction potential (V_{red}) the current generated decreases till point (c). Then, the current starts to decrease when the voltage

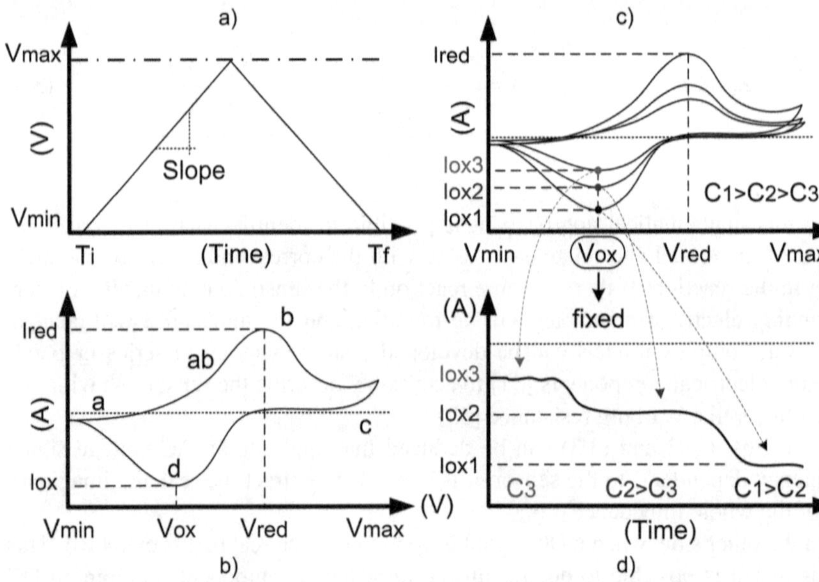

Fig. 3.4 (**a**) Cyclic voltammetry (CV) potential waveform; initial (T_i) and final (T_f) time and maximum (V_{max}) and minimum (V_{min}) voltage, (**b**) CV plot or voltammogram; oxidation and reduction potentials (V_{ox}, V_{red}) and currents (I_{ox}, I_{red}), (**c**) Several CV carried out with the same solution at different concentrations **c** and (**d**) amperogram; fixing a DC potential measure the current

is reversed because the electrons are forced to move from the substance to the electrode, defining the reversible Redox reaction. The minimum current is achieved when the applied voltage is the oxidation potential (d).

The current generated is proportional to the concentration of the analyzed substance and the size of the electrodes [38]. So, if the same analysis with the same type of sensors is carried out in the same substance at two different concentrations, the generated reduction and oxidation currents are different as it is showed in Fig. 3.4c.

At that point, the conception of the *Amperometric* analysis could be introduced. A DC amperometric configuration consists in applying a constant voltage at the $R - W$ electrodes and measuring the current at the *Working* electrode.

Generally, the applied DC voltage is a potential where the electrochemical reaction is maximized. Then, the substance concentration could be continuously monitored and represented in an amperogram. Fig. 3.4d presents the idea of the amperogram using the Oxidation voltage as a fixed potential.

Electrochemical Impedance Spectroscopy (EIS) technique consists in applying an AC voltage to the $R - W$ electrodes and measuring the resulting AC current at the *Working* electrode. Then, it is possible to represent the impedance, or capacitance, variations of the analyzed substance dividing the AC voltage by the current as it is described in Eqs. (3.2) and (3.3).

The typical representation of the EIS is the Nyquist diagram, where the imaginary part of the impedance is plotted against the real one for several frequencies (from low to high frequencies) [36].

Figure 3.5 presents a Nyquist plot for the ideal model introduced in Fig. 3.3. If there is a modification in the concentration of the target substance, the total

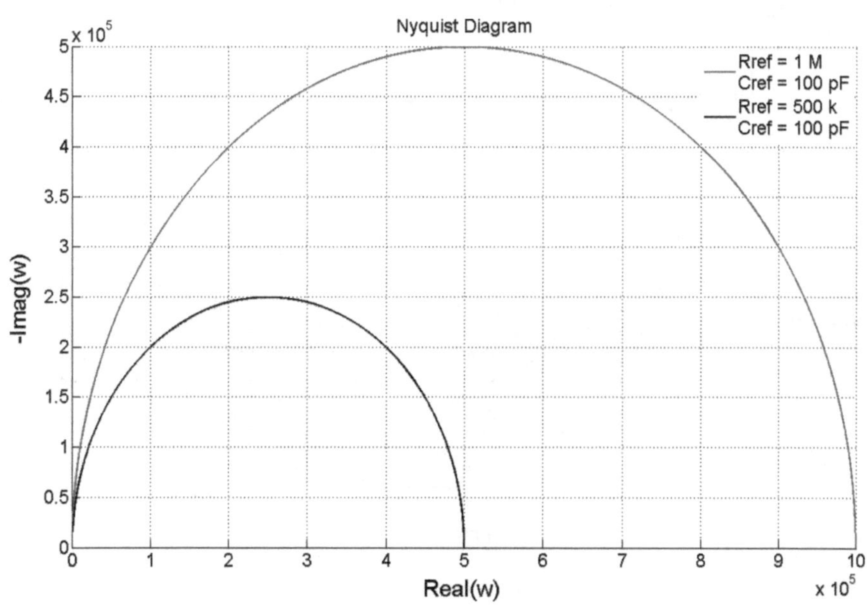

Fig. 3.5 Nyquist plot derived from computer generated data using Eq. (3.3)

impedance of the sensor is also modified as well as the Nyquist plot (i.e. in capacitive sensors the impedance is modified in function of the sensor capacitance, which varies in accordance with the amount of substance detected). Moreover, there are other types of EIS representation like bode plots, where the real and imaginary values are plotted in function of the frequency [24].

Similar to the previous concept of DC amperometric analysis there is the AC amperometry. It consists in using an AC signal at one defined frequency, where the sensor is more sensitive to impedance variations. In that way, the real and imaginary part of the impedance could be continuously controlled at the defined frequency by simply using a specific electronic circuit.

3.3 Potentiostat (Sensor Instrumentation)

The Potentiostat Amplifier (PA) [39, 40, 41] is the name of the electronic instrumentation in charge to control the three Biosensor electrodes. Basically, the PA is formed by two main blocks: Control (CM) and Transimpedance Modules (TIA), Fig. 3.6. The CM keeps the difference voltage between the *Reference* (R) and *Working* (W) electrodes constant by following a well-known input signal (V_{in}). On the other hand, the Detection Module (TIA) measures the current flowing through the W and it also provides a virtual ground to the *Working* electrode.

In the adopted implemented architecture, the Control Module is based on a comparator configuration and a buffer stage, whereas the Detection Module is formed by a Current – to – Voltage converter (TIA). The selected topology is depicted in

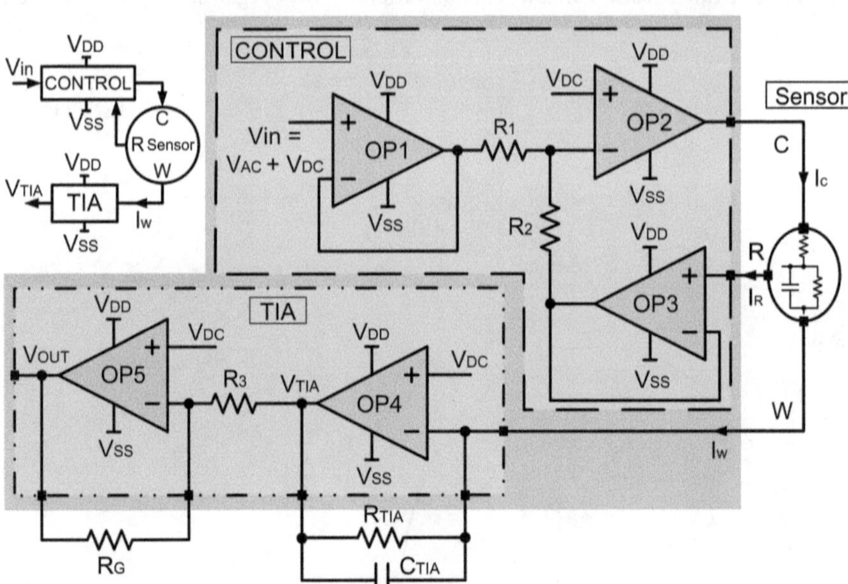

Fig. 3.6 Potentiostat architecture, generic *block diagram* (*white*) and implemented structure (*grey*)

Table 3.1 Potentiostat expected features

Symbol	Parameter	Min	Max	Units
P_{SUPPLY}	Power consumption	< 200 μ		W
V_{SUPPLY}	Voltage supply	±0.6 (1.2)	±1.2 (2.5)	V
V_{in}	Input voltage to apply at V_W-V_R	±10 m	±1	V
Freq.	Frequency at V_W-V_R	DC	10 k	Hz
I_{DET}	Current conversion (I_W)	1 n	100 μ	A

Fig. 3.6 and it is formed by five amplifiers, OP_1 to OP_5, and three resistors, R_1 to R_3. R_{TIA} and R_G are off-chip resistors [42, 43].

The designed potentiostat is focused on Low-Voltage Low-Power applications, like integrated in-vivo monitoring or detection [44], and it is designed to work with low-frequency integrated electrochemical Biosensors [29, 45, 46]. Moreover, the selected topology presents a good trade-off in terms of design simplicity and performance reliability. Table 3.1 summarizes the desired features for the PA.

3.3.1 Control Module (CM)

The Control Module is the specific part of the potentiostat designed to manage the voltage applied to the sensor. This circuitry is in charge to apply and to control the input voltage (V_{IN}) between the *Working* (W) and *Reference* (R) electrodes, such that, $V_{in} = V_W - V_R$ where, V_R is the voltage at the *Reference* electrode and V_W at the *Working* electrode. The adopted CM is based on three operational amplifiers, OP1, 2, and 3, Fig. 3.6.

OP3 monitors continuously the potential (V_R) at the *Reference* electrode, in that way, it is possible to control the voltage drop between W and R. Furthermore, OP 3 is used to ensure a minimal current flow through the R electrode (ideally zero) so, it is possible to assume that:

$$\left.\begin{array}{r} I_C = I_R + I_W \\ I_R \cong 0 \end{array}\right\} \rightarrow I_C = I_W \tag{3.4}$$

where I_C is the current flowing from the Counter or *Auxiliary* electrode, I_R the current in the *Reference* electrode and I_W the current flowing through the *Working* electrode. The Reference voltage (V_R) is then compared with the input voltage (V_{in}) by OP2 defining the voltage applied at the reference electrode:

$$\left.\begin{array}{r} \dfrac{V_{in} - V_{DC}}{R_1} = \dfrac{V_{DC} - V_R}{R_2} \\ R_1 = R_2 \end{array}\right\} \rightarrow \left.\begin{array}{c} V_{in} = -V_R + 2V_{DC} \\ V_{DC} = 0 \end{array}\right\} \rightarrow V_{in} = -V_R \tag{3.5}$$

Since $V_W = 0$ thanks to a virtual ground connection (Section 3.3.2), the cell voltage V_{Cell} is given by,

$$\left.\begin{array}{r} V_{Cell} = V_W - V_R \\ V_W \cong 0 \end{array}\right\} V_{Cell} = -V_R \tag{3.6}$$

Using this CM architecture it is possible to easily control the cell voltage and, hence, to control the electrochemical reaction. Finally, OP1 isolates the input voltage from the rest of the circuit. Regarding resistors R1 and R2, it is important to take into account their values. Small values increase the consumption of the potentiostat. Otherwise, the use of high resistance values could produce noise [47] and operational problems. Moreover, a careful layout between them is a must in order to avoid mismatch problems and to reduce the final size of the integrated resistors.

From the previous explanation, it is deduced that amplifiers OP1, OP2 and OP3 have to be designed taking into account some considerations. OP2 requires a high output current capability to drive properly the *Auxiliar* electrode, whereas OP3 requires a low input bias current in order to guarantee a minimum current flowing in the *Reference* electrode. The maximum operation frequency is also another important aspect to evaluate. If the desirable working frequency of the potentiostat is 10 kHz (Table 3.1) this implies a unity frequency of, at least, 1 MHz for each amplifier [47–49].

Figure 3.7 shows the selected architecture of the amplifier designed to accomplish all the design constrains introduced on Table 3.2. It is composed by a differential PMOS input stage with a folded cascode current multiplier stage followed by an AB output buffer. This topology presents a high output current and slew rate capability for OP2 and low input bias current around 600 pA, enough for OP3.

A differential PMOS input pair is used to assure a good response for low voltages closed to 0 V [50], whereas a folded cascode current multiplier is implemented to achieve enough gain and stability. Finally, a class AB output is selected as output stage for its high current output capability, high slew rate and reduced power consumption.

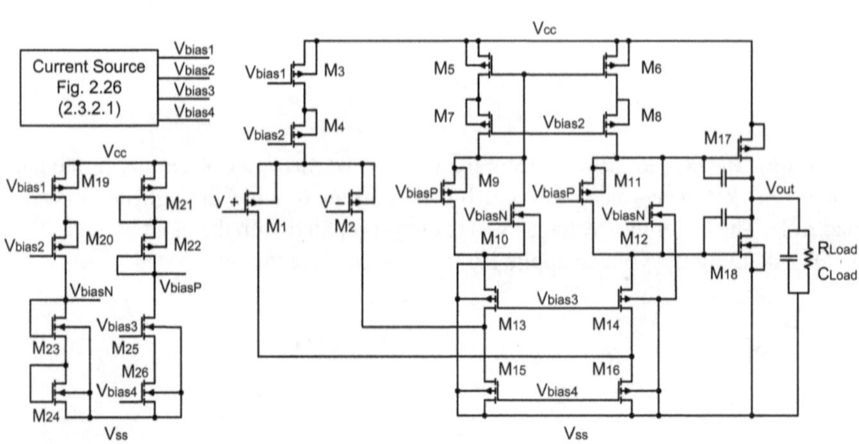

Fig. 3.7 Schematic of the class AB operational amplifier for OP1, OP2 and OP3

Table 3.2 Control module operational amplifiers design constrains

Amplifier	Constrains
OP1; OP2; Op3	↑ Unity frequency (f_u) around 1 MHz
OP1; OP2; Op3	↓ Minimum input voltages
OP2	↑ Output current
OP3	↓ Input bias current

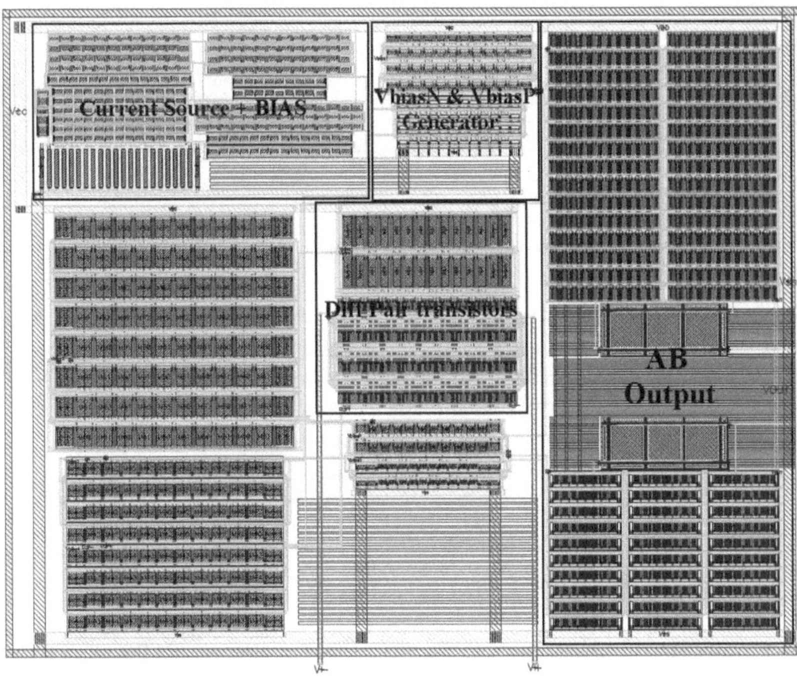

Fig. 3.8 2.5 V AB operational amplifier layout (445 μm × 656 μm)

Two different layouts have been designed. The first one focuses on working with 2.5 V (±1.2 V) supply voltage and optimizing the matching using symmetrically structures, Fig. 3.8. Its main drawback is the huge size of its layout around 445 μm × 565 μm. The second layout prepared to work with a supply voltage up to 1.2 V (±0.6 V) is also implemented. Its layout structure is similar to those shown in Fig. 2.25 involving a total size of 180 μm × 50 μm, which is ideal for small, portable and low-power System on Chip. Table 3.3 summarizes the experimental electrical feature for each amplifier, 2.5 and 1.2 V voltage supply.

At this stage let us analyze the stability of the Control Module. It is important to assure a perfect stability for all load conditions range (sensors) that the potentiostat could work. In that way, problems regarding wrong measurements, wrong

Table 3.3 Class AB amplifier experimental features ($R_{Load} = 1\ M\Omega$, $C_{Load} = 20$ pF)

Symbol	Parameter	Value		Units
V_{CC}	Power voltage	1.2 (±0.6)	2.5 (±1.2)	V
V_{Offset}	Offset voltage	–15.84	–7.92	μV
F_U	Unity frequency	448.88	632.17	kHz
Ph	Phase margin	59.32	61.19	°
G_{DC}	DC Gain	105.64	115.13	dB
V_{noise}	Noise	16.41	14.89	μV
V_{noise}	Noise	–95.7	–96.54	Db
I_{CC}	Current consumption	57.462	74.696	nA
P_{CC}	Power consumption	68.655	186.74	μW
SR @ 10 kHz	Slew rate	0.1927	0.2608	V/μs
SR @ 10 kHz	Slew rate	0.1783	0.2971	V/μs
Size	W × L	180 × 50	445 × 565	μm × μm

sensor polarization or even extra heating dissipation in the AB operational or in the potentiostat could be solved.

Basically, the stability involves resistances, $R_1 - R_2$ operational OP1, 2 and 3, and the *Auxiliar* and *Reference* electrodes. It is also assumed a perfect grounded connection of the *Working* electrode through the Transimpedance Amplifier OP4 (Section 3.3.2). Hence, there is no influence coming from this part of the circuit to the whole stability. Moreover, voltage follower amplifiers OP1 and OP3 can be treated as a short circuit connection in order to reduce the complexity of the analysis.

3.3.2 Detection Module (I-to-V Conversion)

The Detection Module, as the name indicates, is in charge to detect the current flowing in the *Working* electrode, which is the current generated by the electrochemical reaction. The implemented solution is a Current – to – Voltage converter based on a Transimpedance Amplifier (TIA).

Basically, the Transimpedance Amplifier has two main functions: the first one, as it is commented above, consists on measuring the current in the W electrode; whereas the second function consists in providing a virtual ground to the *Working* electrode in order to guarantee Eq. (3.6) and hence, the correct operation of the potentiostat.

The Detection Module based on a TIA is formed by OP4 and the programmable feedback resistance R_{TIA} as it is depicted in Fig. 3.6. The current – to – voltage conversion is done such that,

$$\left.\begin{aligned} V_{TIA} &= -I_W R_{TIA} \\ V_{OUT} &= -V_{TIA}\frac{R_G}{R_3} \end{aligned}\right\} \rightarrow V_{OUT} = I_W R_{TIA}\frac{R_G}{R_3} \tag{3.7}$$

where I_W is the current through the *Working* electrode. Then, combining Eqs. (3.2) and (3.7) and taking into account the virtual ground connection in the W electrode, it can be demonstrated that the Cell Impedance (Z_{CELL}) is described by:

$$Z_{Cell}(jw) = Z_{W-R}(jw) = \frac{V_W - V_R}{I_W}(jw) = \frac{V_R}{V_{OUT}}(jw)\frac{R_{TIA}R_G}{R_3} \quad (3.8)$$

In that way, combining the Control Module and the TIA, it is possible to detect continuously the current, Eq. (3.7), and the impedance Eq. (3.8) of the biochemical interaction. Voltammograms and DC amperograms could be represented directly by using the proportional voltage V_{OUT}, whereas for impedance detection an extra circuit is needed to find out the real and imaginary part (Section 3.4).

At the input of the TIA amplifier a high input resistance is necessary in order to generate properly the virtual ground and the current conversion. If the resistance is not high enough two drawbacks could appear: first, I-to-V conversion errors due to high input bias currents and second, noise problems at the *Working* electrode modifying the virtual ground connection. To avoid those problems, JFET transistors are the perfect choice to be used in the input pair thanks to its high gate input resistance ($\approx$ 10 GΩ) [51, 52]. Otherwise, extra specific input stages are necessary to increase the input resistance if only MOSFET transistors are available.

The final TIA architecture is presented in Fig. 3.9. It is based exclusively on MOSFET transistors. It is formed by a modified input stage to generate a high input resistance ($\approx$ GΩ) and a folded cascode as a second stage. Its input stage is composed by a PMOS differential transconductance pair (M1 M2) with small width transistors to increase the input resistance [53]. Then, an extra differential PMOS

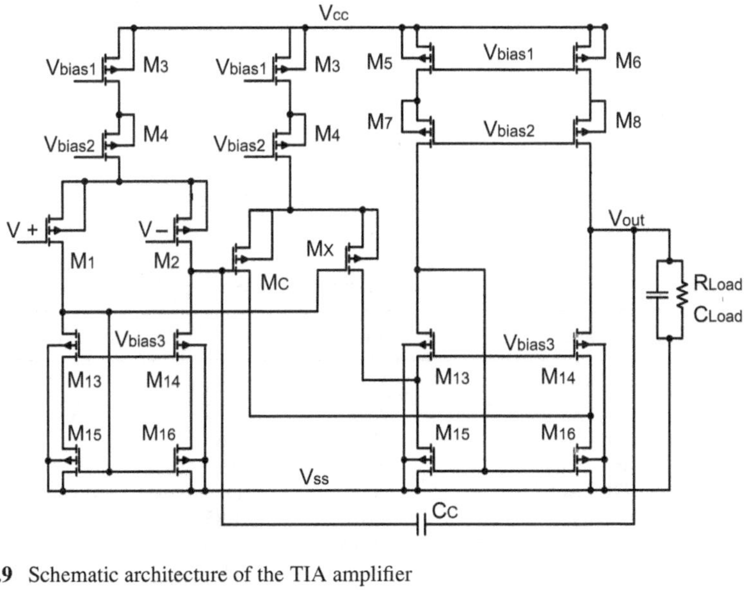

Fig. 3.9 Schematic architecture of the TIA amplifier

pair (MX MC) is added to increase both, the input resistance and also the gain of the whole input stage.

Finally, a folded cascode is implemented as a second and output stage. This circuit increases the input resistance as well as the total gain. Moreover, it isolates the input and output stages while increasing the total bandwidth [53]. An extra capacitor C_C (5 pF) between the input and the output assures the stability for the whole frequency range.

Since the output stage is an ordinary folded cascode, it is not possible to drive enough current at the output, so it must be followed by an extra Gain amplifier (OP5), which is based on the class AB presented in Fig. 3.7.

Following the same idea introduced in the Control Module, two different layouts have been designed. The first one focuses on working with 2.5 V and on optimizing the matching using symmetrically structures, similar to Fig. 3.8. Its main drawback also concerns its size of 280 μm × 390 μm.

The second layout prepared to work with a supply voltage up to 1.2 V (±0.6 V) is also made. Its layout structure is similar to the one shown in Fig. 2.25 a total area of 180 μm × 50 μm ideal for small, portable and low power Systems on Chips. Table 3.4 summarizes the experimental electrical feature for each TIA.

A feedback capacitor C_{TIA}, Fig. 3.6, could be added in parallel with R_{TIA} in order to control the frequency response of the Transimpedance Amplifier and to define a frequency filter [47, 54]. The cut-off frequency could be easily calculated using Eq. (3.9) for a simple filter response.

$$f^{TIA}_{cut-off} = \frac{1}{(2\pi R_{TIA} C_{TIA})} \tag{3.9}$$

The previous filter approach does not take into account the impedance of the Sensor Cell. Hence, other type of filter responses could be adapted for more accurate frequency restrictions. A 2nd order Butterworth [55] response is obtained following Eq. (3.10):

$$\frac{1}{(2\pi R_{TIA} C_{TIA})} = \sqrt{\frac{GBP}{(4\pi R_{TIA} Z_{Cell})}} \tag{3.10}$$

Table 3.4 TIA experimental electrical features $R_{Load} = 1\ M\Omega$, $C_{Load} = 20$ pF

Symbol	Parameter	Value		Units
V_{CC}	Power voltage	1.2 (±0.6)	2.5 (±1.2)	V
V_{Offset}	Offset voltage	23.1	14.36	μV
F_U	Unity frequency	398.68	570.17	kHz
Ph	Phase margin	54.23	63.9	°
G_{DC}	DC Gain	70.57	85	dB
R_{IN}	Input resistance	1.02	1.85	GΩ
I_{CC}	Current consumption	30.25	51	nA
P_{CC}	Power consumption	35.41	118.64	μW
Size	W × L	180 × 50	280 × 390	μm × μm

where GPW is the gain bandwidth product of the TIA and the Z_{Cell} is the impedance between *Working* and *Reference* electrodes. In that implementation, it is important to know an approximate value of the Z_{Cell} range to design properly the cut-off frequency, Eq. (3.11).

$$f_{cut-off}^{TIA-But} = \sqrt{\frac{GBP}{(4\pi R_{TIA} Z_{Cell})}} \tag{3.11}$$

3.3.3 Results

Two potentiostats are designed in a commercial 0.13 μm technology (Appendix B) from ST Microelectronics. One of them is focused on low voltage operation up to ±0.6 V and low area consumption. The second one, as it is commented previously, is focused on working with a voltage up to ±1.2 V and on improving the layout matching. Figure 3.10 depicts the BIO-Chip with the two potentiostats as well as other circuits.

Several tests have been carried out in order to validate the Control and Detection modules of the potentiostat. First of all, the linearity and the PSRR of the potentiostat have been studied using some Randles models.

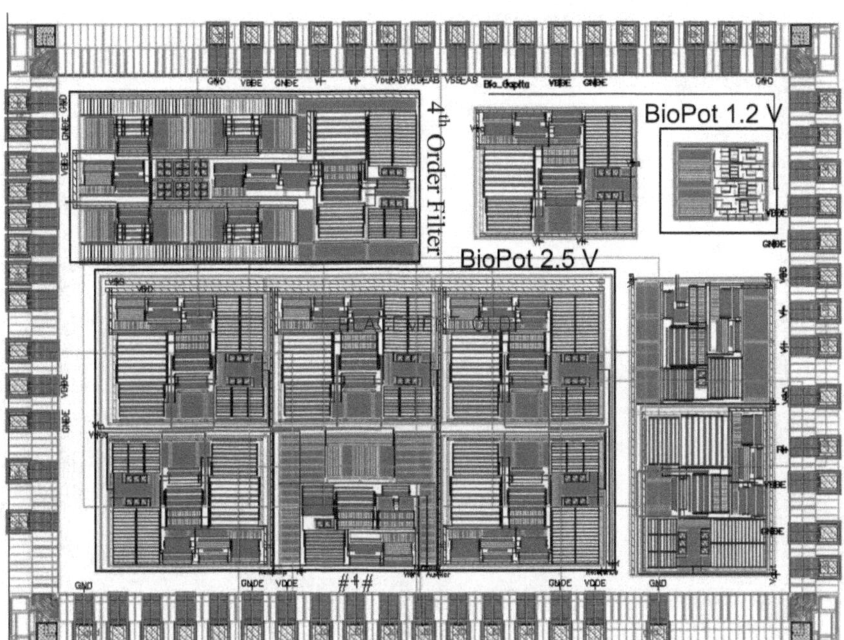

Fig. 3.10 Layout image of the IC "bio-chip" with the two designed potentiostats and the OTA filter

Then, the performances of the potentiostat are studied using several concentrations of potassium ferrocyanide, $K_4[Fe(CN)]_6$ [56] and a real sensor. Cyclic voltammograms and amperometric analysis are carried out and the results are compared with the ones obtained using a commercial instrument from CH Instruments [57].

The BVT Technologies AC1.W1.R1 [58] sensor is the selected option to proceed with the measurements. It is a ceramic based sensor formed by a gold working and auxiliary electrodes and a silver/silver chloride reference electrode. Furthermore, a bio-chemically active substance could be immobilized on the *working* electrode to create a Biosensor. That sensor is mainly focused on the measurements of glucose, ferricyanide, enzyme activity and basic electrochemical and bio-electrochemical techniques.

The test setup is composed by a LabVIEW [56] based software and a full-custom PCB for the IC socket, sensor and other connections. Figure 3.11 shows the PCB with the 68-pins BIO-Chip, the AC1.W1.R1 sensor and a capture of the LabVIEW main window; on it, a voltammogram and an amperometric analysis

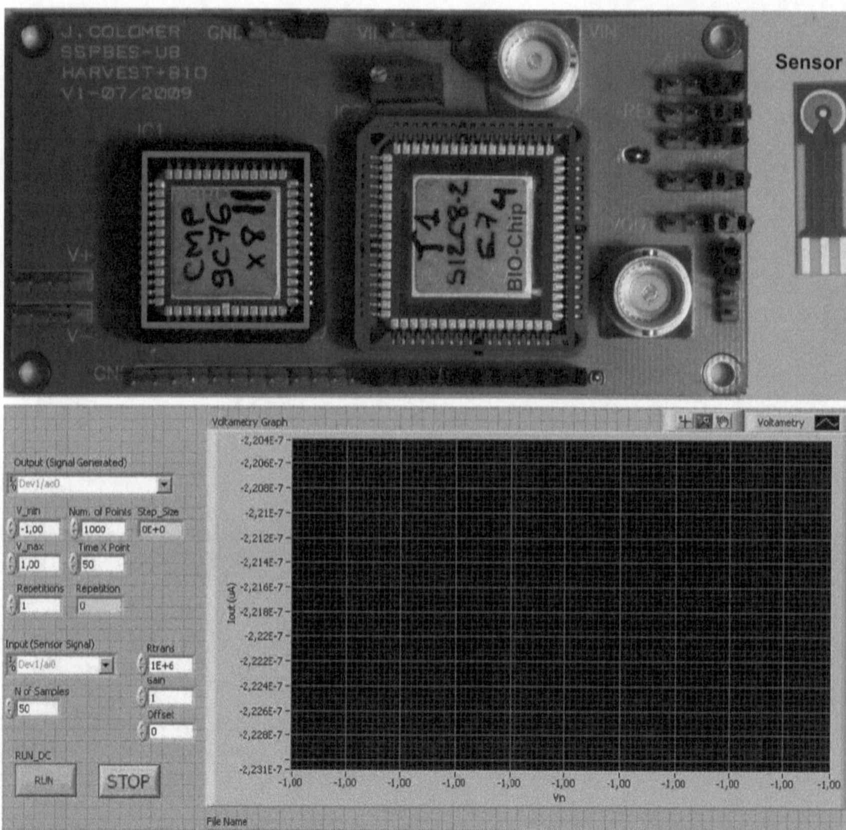

Fig. 3.11 PCB bio-chip test board with sensor connections and AC1.W1.R1 test sensor (*top*) and lab-view acquisition program main window (*bottom*) [55]

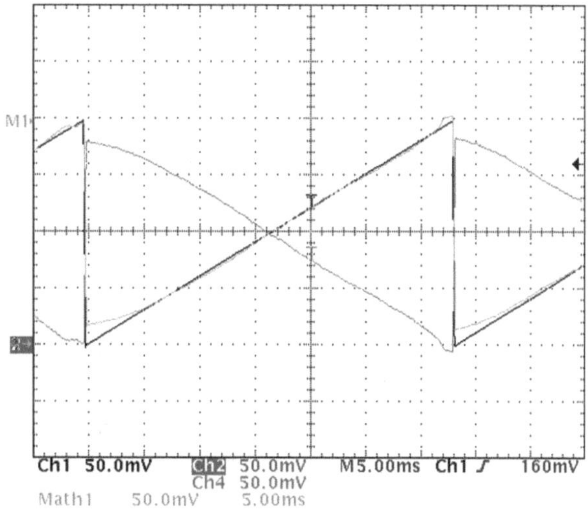

Fig. 3.12 Linear range of the 1.2 V potentiostat for an input ramp of 100 mV

could be realized (not at the same time). The results are displayed on the screen and stored in a plain file for post processing.

In order to represent the sensor cell, a Randles circuit formed by the double-layer capacitor (C_{ref} = 100 pF), in parallel with a polarization resistor (R_{ref} = 100 kΩ), which is also described as a charge transfer resistor, and the solution resistor (R_{aux} = 50 kΩ), has been used, Fig. 3.3.

The linearity of the potentiostat amplifier has been experimentally validated. The DC linear range for the 1.2 V (±0.6 V) potentiostat is ±0.54 V, whereas for the 2.5 V (±1.2 V) potentiostat is ±1.12 V with a deviation of less than 1% for both of them. Taking into account that small DC voltages could be applied for in vivo applications, the designed potentiostats have a correct range. Figure 3.12 depicts the cell voltage of the 1.2 V potentiostat for an input ramp up to 100 mV.

Experimental transient tests have been also carried out to analyze the AC performances and how the V_{Cell} follows an AC V_{in} signal. Figurer 3.13 shows the Working, the input and the TIA output voltages obtained with the 1.2 V potentiostat.

The Randles model uses a R_{aux} = 3.9 MΩ, R_{ref} = 1 MΩ and C_{ref} = 100 pF. The transimpedance resistance (R_{TIA}) is fixed at 1 MΩ. A sinus input signal is introduced with a DC component of 50 mV and an AC component of 100 mV_{PP} and a frequency of 1 kHz. The Total Harmonic Distortion (THD) is around 2%.

The frequency response of both potentiostats has also been studied. A fixed Randle model with R_{aux} = 3.9 MΩ, R_{ref} = 1 MΩ and C_{ref} = 100 pF allows defining a cut-off frequency (F_{-3db}) around 1,500 Hz. The ideal frequency response of the Cell Impedance is represented in Fig. 3.14.

Then, the experimental Z_{Cell} is obtained using Eq. (3.8) with $R_{TIA} = R_G = R_3 =$ 1 MΩ. Both potentiostats have a maximum working frequency lower than 10 kHz, 2,200 Hz for 1.2 V potentiostat and 8,700 Hz for 2.5 V potentiostat. The 1.2 V potentiostat experimental responses could be compared with the theoretical ones in Fig. 3.14. There, it could be observed how the response of the potentiostat saturates in its maximum working frequency.

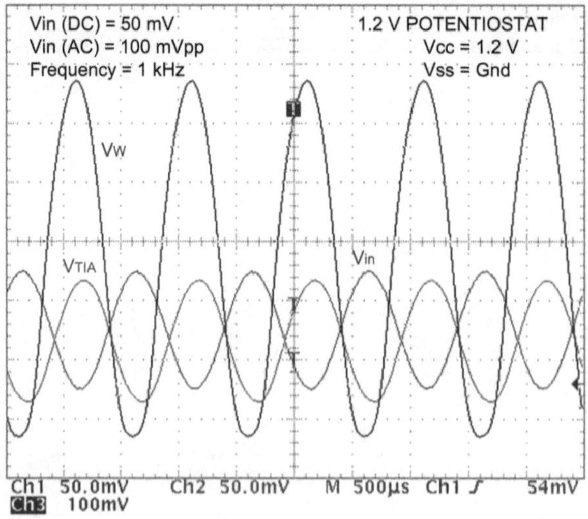

Fig. 3.13 1.2 V potentiostat AC response. V_W is the voltage at the working electrode. V_{in} is the input voltage and V_{TIA} is the voltage at the output of the transimpedance amplifier. A Randles model formed by $R_{aux} = 3.9\ M\Omega$, $R_{ref} = 1\ M\Omega$ and $C_{ref} = 100$ pF and a transimpedance resistance of 1 MΩ are used

Finally, both potentiostats have been calibrated before using them with real sensors. The calibration is done using several values of R_{ref} (470 kΩ, 1.2 MΩ, 1.5 MΩ, 5.6 MΩ) and applying a DC sweep between *Working* and *Reference* electrodes. Moreover, the R_{aux} resistance of the Randles model is replaced by a short

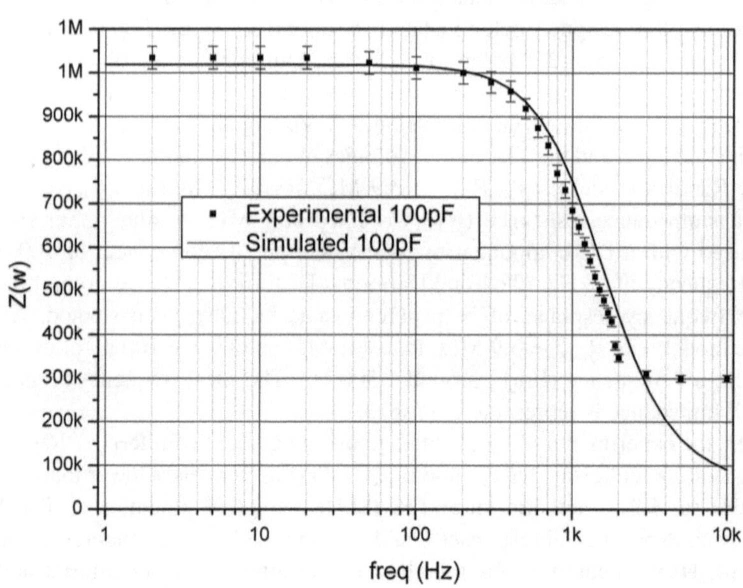

Fig. 3.14 1.2 V potentiostat frequency response

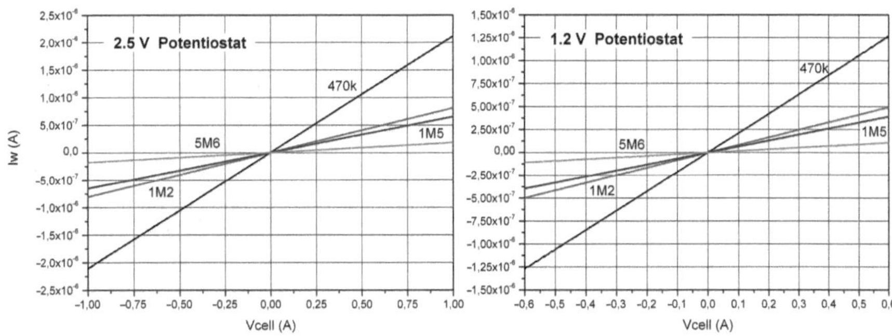

Fig. 3.15 Calibration curves for 1.2 and 2.5 V potentiostats for different R_{ref} values

circuit between *Counter* and *Reference* electrodes and the transimpedance (R_{TIA}) resistance is fixed at 1 MΩ in both cases.

The procedure is simple: a DC sweep is performed for each calibration resistance while reading the output voltage of the TIA. Then, the current ($I_W = V_{Cell}/R_{ref} = V_{TIA}/R_{TIA}$) is represented in function of the applied DC voltage (V_{Cell}), Fig. 3.15. The waveforms show that both potentiostats detect the correct current value for all voltages so, they are calibrated.

The last step consists in performing two electrochemical analysis, cyclic voltammetry and amperometric analysis, using five concentrations of $K_4[Fe(CN)]_6$ (1, 2, 3, 4 and 5 mM) in order to validate the designed integrated potentiostats. The results are compared with the ones obtained with the commercial instrument CHI 1232A from CH Instruments [57] (Appendix C shows some specifications).

The AC1.W1.R1 [58] sensor is connected to the potentiostats and a Cyclic Voltammetry (CV) from –1 to 1 V at 0.05 V/s is carried out with the 2.5 V potentiostat and the CH 1232A for all concentrations. A CV from –0.6 to 0.6 V at 0.05 V/s is done for the 1.2 V potentiostat. The results of each voltammetry are depicted in Fig. 3.16.

The shape of the electrochemical reaction is almost equal in all three potentiostats. The reduction and oxidation potentials are located around the same voltages of 0.170 V and 0.240 V for all potentiostats as can be observed in charts (d), (e), (f), (g) and (h).

The main difference lies in the current peak amplitude of the oxidation and reduction points. A 2.5 V integrated potentiostat presents maximum difference of 8 μA with respect to the commercial one. On the other hand, the 1.2 V potentiostat presents a higher difference around 12 μA; this could be caused by its lower input impedance than the 2.5 V potentiostat in the Transimpedance amplifier.

However, part of these differences is due to reproducibility errors of the electrochemical measure; it is not possible to obtain the same exact result every time it is performed the same experiment. The obtained CV waveforms are similar to those that can be found in the literature like [33, 35, 59–61].

The last electrochemical analysis consists on performing an amperometric analysis of several concentrations of $K_4[Fe(CN)]_6$ for the 2.5 and 1.2 V integrated potentiostats and then compare the results with those obtained with the commercial instrument.

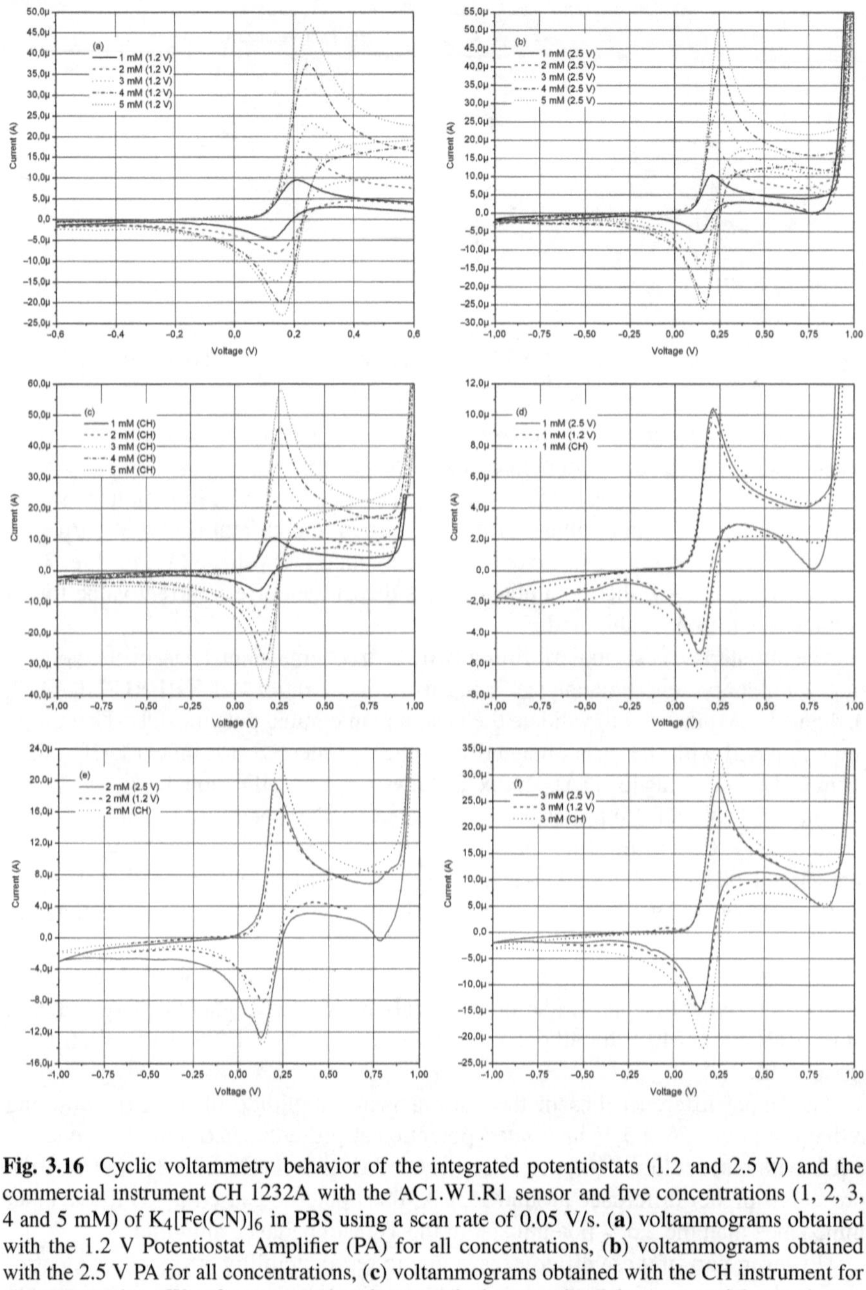

Fig. 3.16 Cyclic voltammetry behavior of the integrated potentiostats (1.2 and 2.5 V) and the commercial instrument CH 1232A with the AC1.W1.R1 sensor and five concentrations (1, 2, 3, 4 and 5 mM) of $K_4[Fe(CN)]_6$ in PBS using a scan rate of 0.05 V/s. (**a**) voltammograms obtained with the 1.2 V Potentiostat Amplifier (PA) for all concentrations, (**b**) voltammograms obtained with the 2.5 V PA for all concentrations, (**c**) voltammograms obtained with the CH instrument for all concentrations. Waveform comparison between the integrated and the commercial potentiostats for all concentrations: (**d**) 1 mM, (**e**) 2 mM, (**f**) 3 mM, (**g**) 4 mM and (**h**) 5 mM

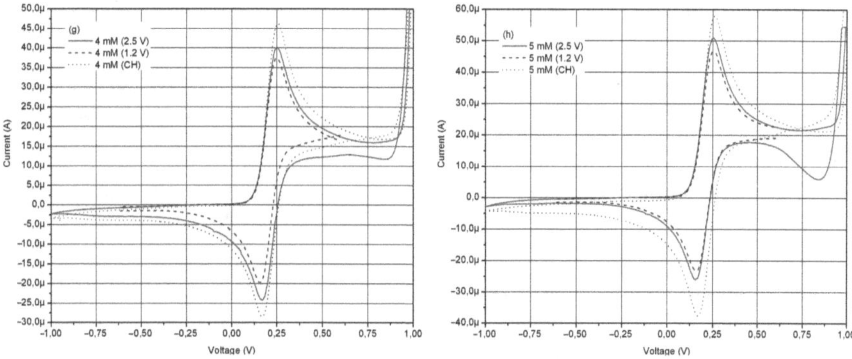

Fig. 3.16 (continued)

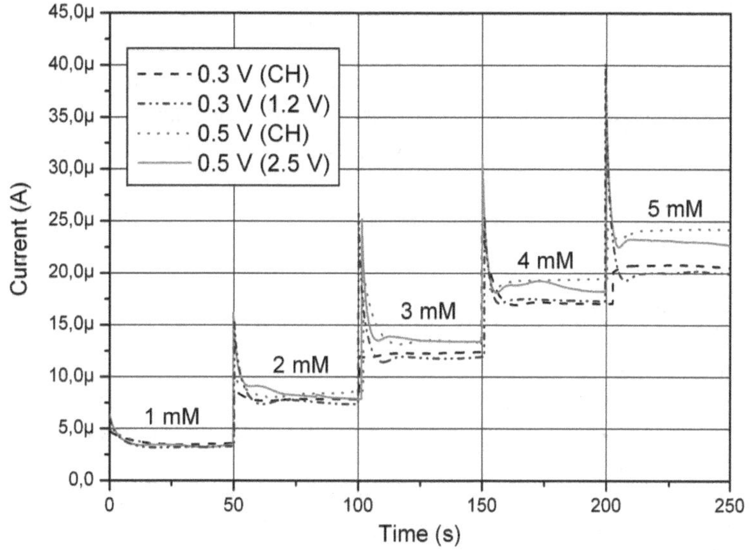

Fig. 3.17 Amperogram recorded during the electrochemical reaction of 1, 2, 3, 4 and 5 mM of $K_4[Fe(CN)]_6$ in PBS for the three potentiostats. The applied cell voltage is 0.5 V for the 2.5 V integrated potentiostat and 0.3 V for the 1.2 V potentiostat. The amperometries are compared with the ones obtained by the commercial instrument CH 1232A (CH)

For the 2.5 V integrated potentiostat a potential of 0.5 V is applied to the amperometry whereas for the 1.2 V potentiostat a voltage of 0.3 V is used. The obtained amperograms are presented in Fig. 3.17 where can be compared with the ones obtained with the CH 1232A. In that case, the results are similar between the commercial and the full-custom integrated potentiostats. The current peaks are produced when the new concentration is added to the solution. They can be reduced using a low pass filter in order to avoid them in the design of some applications.

Finally, Table 3.5 summarizes the hardware specification for each designed integrated potentiostat.

Table 3.5 2.5 and 1.2 V potentiostats hardware specifications

Symbol	Parameter	Value		Units
V_{CC}	Power voltage	1.2 (±0.6)	2.5 (±1.2)	V
Potential[a]	V_{Cell} voltage	1 (±0.5)	2 (±1)	V
Max. current	Max. current detected	±1	±1	mA
Min. current	Min. current detected	±1	±0.1	nA
Freq.	Max. working freq.	2,000	8,000	Hz
P_{CC}	Power consumption	240	580	μW
Size	W × L	1,700 × 900	300 × 250	μm × μm

[a]Voltage across *Reference* and *Working* electrodes

3.4 Low-Frequency Lock-In Amplifier

As previously commented, in some cases the sensor gives the information about the electrochemical reaction by modification on its impedance [28]. Not a full Electromechanical Impedance Spectroscopy (EIS) is carried out and a variation of the Biosensor impedance (or capacitance) is just sensed at one or various defined frequencies where the sensor is more sensitive to changes on the analyte concentration [62]. In those cases, a specific electronics able to evaluate the complex value of impedance is necessary.

Label free impedance Biosensors can detect a great variety of targets, like proteins, DNA, small molecule sensors [63], cell-based Biosensors [62, 64], etc. Due to their low cost, low power and ease of miniaturization, electrical Biosensors hold great promise for applications where minimizing size and cost is crucial, such as point-of-care diagnostics, biological warfare agent detection and integrated in-vivo point-of-care detectors.

The signals to be measured have often very small amplitude and they are usually in a very noisy environment. Then, it is mandatory to extract the signal from the noise and a very promising approach is based on the use of a lock-in amplifier [65] capable to identify the real and imaginary parts of the complex impedance while reducing the effect of the noise [66, 67].

Looking for this kind of implementations, there is a trade-off between complexity, area, power consumption, with the desired measurements and the electronic implementation. In that way, a fully-integrated DSP solution, as a digital lock-in amplifier, would present a big challenge to develop an implantable low power device. Obviously, commercial lock-in amplifiers are costly and large, and they are not suitable for portable electronics.

A miniaturized solution based on an analog lock-in amplifier module has to be designed in order to measure impedances associated with the analysis of biological cells, in the frequency range of interest. A digital solution is not developed due to its large silicon area needed in the integrated implementation and its higher power dissipation compared with a full-custom analog solution.

Following this assumption it is designed (schematic and layout level) an analog lock-in amplifier as an electronic interface for implantable low frequency Biosensors applications. The presented design looks for a very small power consumption working at very low power supply (1.2 V).

3.4.1 Lock-In Amplifier Architecture

The impedance detection [68] is carried out by introducing an AC signal to the sensor; hence an AC signal is obtained in the sensor output. In that way, when working with amperometric sensors, an AC output current is obtained and the impedance could be defined as the division between the input AC voltage and the output current.

The input voltage is applied to the three electrodes Biosensors using the control module presented in Section 3.3.1. On the other hand, a transimpedance amplifier converts the output AC current into a proportional voltage as it is detailed in Section 3.3.2. At that point, it is introduced the lock-in circuit to detect the real and imaginary parts of the impedance. Figure 3.18a presents the block diagram structure of the impedance analyzer instrument.

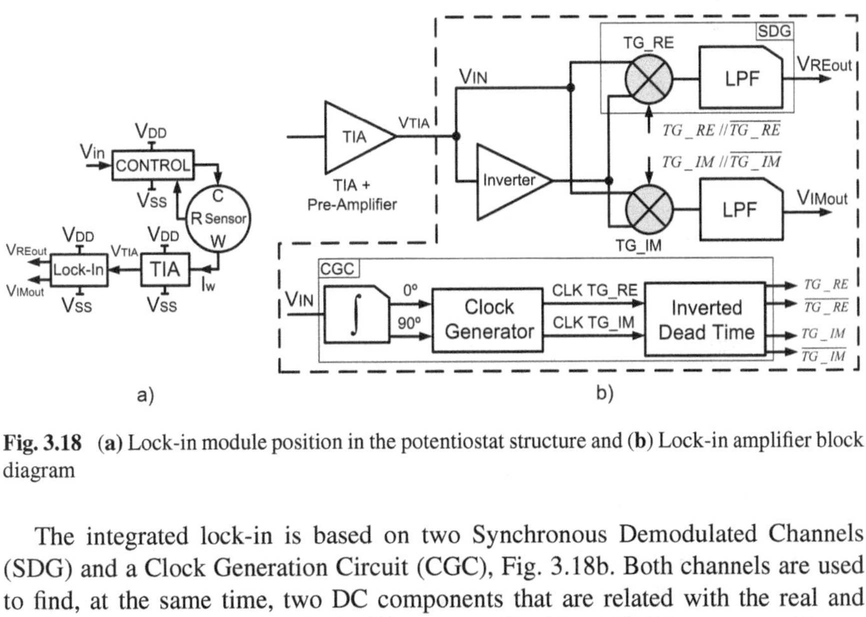

Fig. 3.18 (**a**) Lock-in module position in the potentiostat structure and (**b**) Lock-in amplifier block diagram

The integrated lock-in is based on two Synchronous Demodulated Channels (SDG) and a Clock Generation Circuit (CGC), Fig. 3.18b. Both channels are used to find, at the same time, two DC components that are related with the real and imaginary values related with the Biosensors. The Front-End is composed by an input pre-amplifier used to feed the incoming signal from the I-V converter to the voltage accepted by the channels. On the other hand, the Back-End demodulation is formed by an inverter amplifier, two Low Pass Filters (LPF) and two transmission gates (TG) [69]. This TG demodulator acts as a multiplier.

The adopted Pre-Amplifier configuration is a wide-swing, rail-to-rail output stage with feedforward class-AB control [66, 67] presented in Fig. 3.7 with a

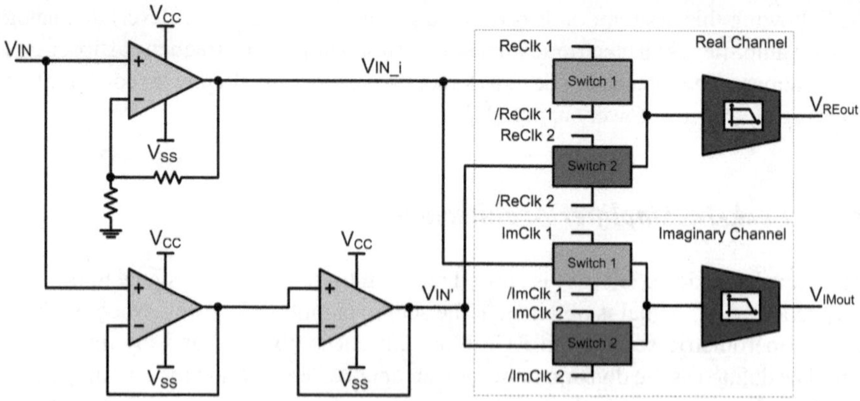

Fig. 3.19 Back-end implementation of the synchronous demodulated channels (SDC)

gain-bandwidth product of 300 kHz for a load capacitor of 20 pF. This pre-amplification is included in the I-V conversion stage.

The demodulator is a multiplier. It takes the input and the reference signals and multiplies them together. When you multiply two waveforms together you get the sum and difference frequencies as a result. Since the input signal and the reference signal are of the same frequency, the difference frequency is zero and you get a DC output which is proportional to the amplitude of the input signal and the cosine of the phase difference between the signals.

By adjusting the phase of the reference signal using the reference circuit, the phase difference between the input signal and the reference can be brought to zero and hence the DC output level from the multiplier is proportional to the input signal.

The Back-end implementation of the Lock-In amplifier is presented in Fig. 3.19 with the inverter and the real and imaginary channels. Furthermore, two voltage followers are included in the non-inverted line to compensate the delay error between channels. Both, the inverter and the follower, use the same previous class AB amplifier.

The Clock Generation Circuit generates the 4 clock signals for the transmission gates (TG) with the same AC signal used to drive the electrochemical cell, Fig. 3.18. Hence, this four clocking scheme defines the on/off state of the analog TG's for a synchronous rectification. Moreover, a Dead Time generation is introduced to reduce the clock feed-through noise [70] as well as, a dummy half size transistor to cancel the charge injection errors.

The first step in the Clock Generation scheme consists on generating a 90% delayed signal from the input ones. This means that the real channel reference clocks are based on a sinus waveform whereas the imaginary channel is controlled by a cosines signal obtained by integration of the original sinus signal.

The 0 and 90° signals are obtained following the architecture depicted in Fig. 3.20. Since a low frequency operation is expected, a classical integration topology based on a class AB operational amplifier is chosen. Its transfer function is

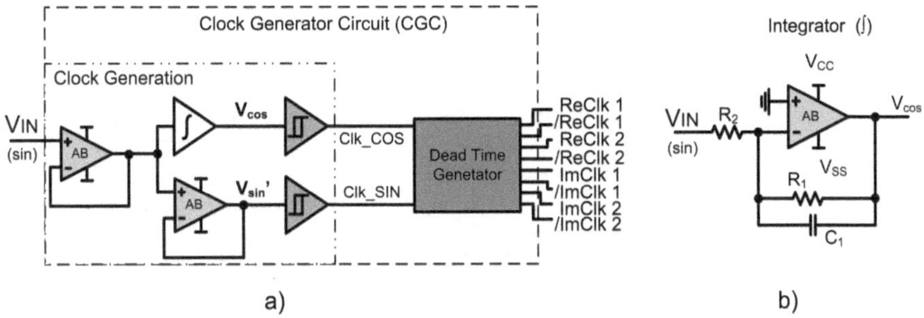

Fig. 3.20 (a) Clock generation circuit architecture and (b) Integrator topology

defined in Eq. (3.9). Moreover, Eq. (3.10) has to be maintained to guaranty the best integration ratio as well as a unitary gain.

$$H(s) = -\frac{R_1/R_2}{sC_1R_1 + 1} \tag{3.12}$$

$$Z_1 = Z_2 \Rightarrow R_2 = (C_1 \| R_1) \tag{3.13}$$

Two important remarks when using this kind of integrators: First, capacitor C_1 has to be pre-charged during the start-up and second, it must be stated that the impedance should be matched so, if the frequency presents a huge range variation, the capacitor C_1 should be changed or other solution must be adopted. Just as in the Back-end, a voltage follower is introduced to compensate the delay produced by the integrator.

Moreover, an extra follower is added at the beginning of the Clock Generator to compensate the delay between the clock generation and the Back-End stage. In that way, all generated on-chip signals present the same delay so, the final error is reduced.

The last step consists on generating the square clock signal from the 0 and 90% sinus waveforms. This is accomplished by using comparators based on the structure presented in Fig. 2.45, Section 2.3.3.2. Figure 3.21 shows the two clock generated signals and the delay between them. It is compensated by adding followers in the Clock Generator circuit and in the Back-End stage.

Later, the two out of phase signals are introduced in the Dead Time Generator to drive properly the CMOS Transmission Gates with four non-overlapped clock signals per channel. Also, dummy half size transistors have been used to cancel the charge injection errors.

Then, the two signals coming from the Transmission Gate rectifiers are filtered through two active low-pass filters (Section 3.4.3) thus, two DC components are obtained, V_{REout} and V_{IMout}, one per each channel. Finally, the magnitude and phase of the impedance are obtained afterwards combining both DC values using Eqs. (3.11) and (3.12).

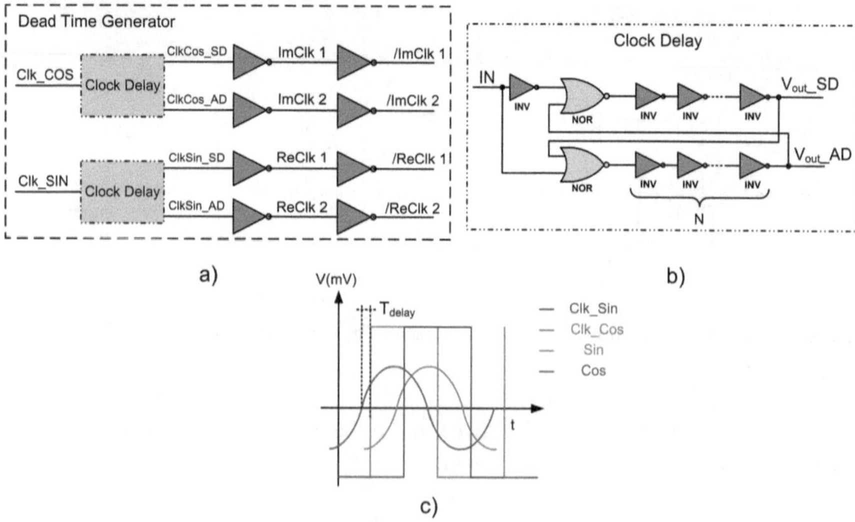

Fig. 3.21 (**a**) Dead time generation structure, (**b**) detailed schematic of the delay generation and (**c**) illustration of delay operation

$$Magnitude = \sqrt{V_{REout}^2 + V_{IMout}^2} \tag{3.14}$$

$$Phase = Tan^{-1}\left(V_{IMout}/V_{REout}\right) \tag{3.15}$$

3.4.2 Lock-In Simulation Results

The performance of the lock-in amplifier is analyzed using the extracted views of the circuitry. An ideal AC signal is used to define the input sinusoidal module and phase (V_{IN}). Fixed amplitude of 60 mV is selected whereas the phase is changed between different values.

Regarding the two DC voltage levels at the output of the filters, V_{REout} and V_{IMout}, the theoretical amplitude and phase are compared with the calculated values using Eqs. (3.11) and (3.12), for a particular frequency of 10 kHz. A maximum error of a 2% is obtained in the Amplitude measurement (A_{ERR}).

Figure 3.22 depicts the Phase simulated values compared with the theoretical ones. In that case, an error around 1% (P_{ERR}) is achieved. Both error values, M_{ERR} and P_{ERR}, are good results taking into account the reported errors by the commercial SR830 lock-in amplifier by Stanford Research Systems©, which presents an Amplitude error (A_{ERR}) around 2.29% @ 10 kHz, as a reference [43].

Table 3.6 presents the simulated performance of the lock-in amplifier when there is some noise applied at the input signal of 60 mV. The table shows the obtained DC

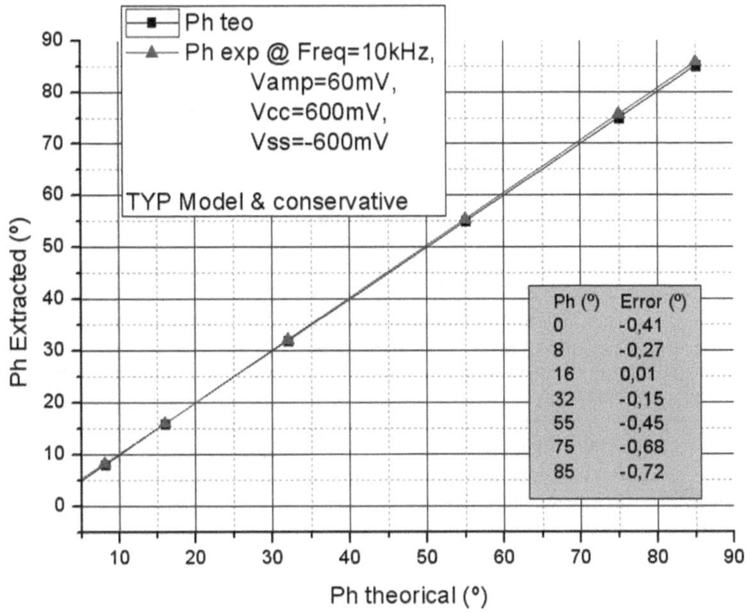

Fig. 3.22 Phase comparison between the calculated (Ph teo) and the simulated (Ph exp) results

Table 3.6 Simulated DC magnitude (V_M) and Phase (φ) obtained with the lock-in amplifier when different noise levels are applied at the input signal of 60 mV @ 10 kHz, V_{cc} = 1.2 V. (φ_{teo} (°) = theoretical phase; V_{noise} = noise amplitude; F_{noise} = noise frequency; $|V_M|$ = magnitude, Eq. (3.11); $\epsilon|V_M|$ = magnitude error; φ_{exp} = phase, Eq. (3.12); $\epsilon\varphi_{exp}$ = phase error)

φ_{teo} (°)	V_{noise} (mV)	F_{noise} (Hz)	$\|V_M\|$ (mV)	$\epsilon\|V_M\|$ (mV)	φ_{exp} (°)	$\epsilon\varphi_{exp}$ (°)
0	5	500	59.89	0.11	1.1	1.1
45	5	500	59.52	0.48	44.79	0.21
0	100	500	59.86	0.14	1.13	1.13
45	100	500	59.2	0.8	44.99	0.01
0	5	15 k	59.88	0.12	1.1	1.1
45	5	15 k	59.06	0.94	44.92	0.08
0	100	15 k	59.98	0.02	1.11	1.11
45	100	15 k	59.04	0.96	45.47	0.47

value of the Magnitude, Eq. (3.11), the Phase, Eq. (3.12), and the error produced in the detection of both magnitudes.

The total power consumption of the lock-in amplifier is 715 μW for a power supply of ±0,6 V. Table 3.7 presents the consumption for different input amplitude voltages and power conditions whereas Table 3.8 summarizes the most important electrical characteristics of the designed lock-in amplifier.

Table 3.7 Lock-in amplifier power consumption in function of several input voltages amplitudes and power supply voltages

Input (10 kHz)		
V_{Supply}(V)	V_{IN}(mV)	Power (W)
±0.6 V	60	714 μ
±0.6 V	100	713 μ
±1.25 V	60	1.2 m
±1.25 V	100	1.2 m

Table 3.8 Lock-in amplifier maximum values

Symbol	Parameter		Value	Units
V_{CC}	Power voltage	1.2 (±0.6)	2.5 (±1.2)	V
V_{IN}	Maximum input voltage	0.6 (±0.3)	1.2 (±0.6)	V
Freq	Maximum input frequency	10	10	kHz
$\epsilon\|V_M\|$	Magnitude error	2	1.7	%
$\epsilon\varphi_{exp}$	Phase error	1.2	1.03	%
P_{CC}	Power consumption	715	1,200	μW

3.4.3 Active Low-Pass Filter

Two active low-pass filters are used as a last step of the previous Lock-In amplifier architecture. They filter any kind of noise and high frequency interferences in order to reduce errors in the obtained DC signals.

Since the integrated filter must present a very low cut-off frequency, ranging from 0.1 to 30 Hz, some classical solutions like Sallen-Key filter could not be adapted because high values of resistances and capacitances are necessary and then, the IC integration is not possible. A gm-C second-order LP filter is adopted to accomplish the frequency requirements and to be able to integrate the necessary capacitors values.

The final implementation uses a degenerated transconductance amplifier (OTA) to define the *gm* of the filter [71]. This architecture allows defining small transconductance values in the range of few nS [47, 54].

Figure 3.23 shows the schematic of the implemented OTA as well as the in/out current relation. The source degeneration increases the input range of the amplifier and also decreases the equivalent transconductance.

The ratio (M:N) between the current mirrors decreases the current level at the output, which results in an even minor value [72]. These current mirrors are based on composite transistors, used to reach greater copy factors.

Important remarks must be followed when designing this type of current mirrors in terms of layout. A better matching is achieved with techniques like common centroid geometry and with multifingers because large numbers of unit transistors are matched together. Figure 3.24 depicts the OTA's layout.

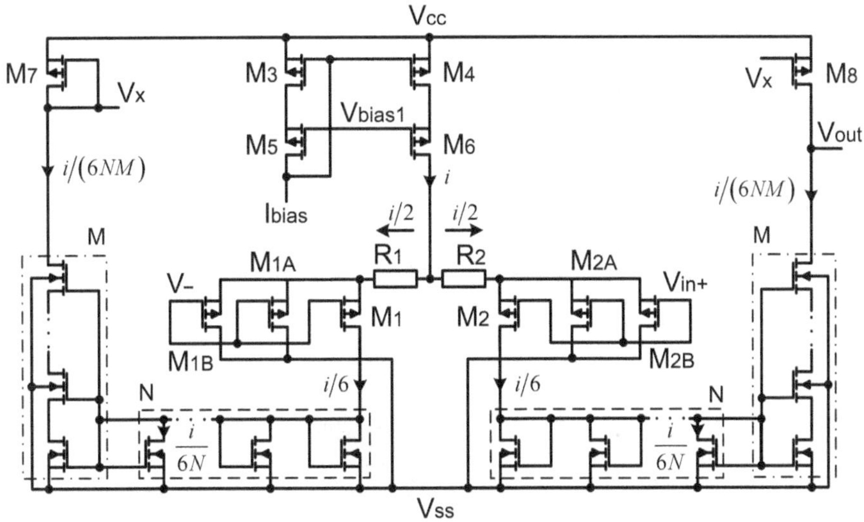

Fig. 3.23 Schematic of the degenerated OTA used to design active filters. Resistors R_1 and R_2 and the M:N ratio between transistors is used to reduce the current at the output stage

Fig. 3.24 Degenerated OTA layout

The equivalent transconductance (gm) for this amplifier is 4.25 nS, for an input range up to ±400 mV and a supply voltage of 1.2 V. The OTA amplifier presents an input offset <1 mV, a THD of 0.03%, and a power consumption around 0.5 μW, for the typical case (TYP Model).

A lower nominal transconductance value up to 1 nS could be achieved. The minimum gm value is limited by the leakage currents off each current mirror branch. Table 3.9 summarizes the OTA electrical specifications.

Table 3.9 Degenerated OTA electrical features

Symbol	Parameter	Value		Units
V_{CC}	Power voltage	1.2 (±0.6)	2.5 (±1.2)	V
V_{Offset}	Offset voltage	770.1	–982.54	μV
V_{in}	Input voltage range	0.8 (0.4)	1.9 (1)	V
F_{3db}	Cut-off frequency	32.95	43.39	Hz
g_m	Transconductance	4.241	4.487	nS
THD	Total harmonic distortion	0.273	0.0073	%
I_{CC}	Current consumption	480.4	570	nA
P_{CC}	Power consumption	582.76	1,124.36	nW
Area	W × L	380 × 186		μm × μm

The basic structure of the 2nd order filter is depicted in Fig. 3.25. It is formed by two OTA amplifiers (*gm1*, *gm2*) and 2 capacitors. Its transfer function is defined by Eq. (3.13) where $V_A = V_{IN}$ and $V_B = V_C = 0$ V whereas the cut-off frequency (ω_O) is described by Eq. (3.14) where $gm_1 = gm_2 = gm_2$.

$$Vout = \frac{s^2 C_1 C_2 V_C + sC_1 gm_2 V_B + gm_1 gm_2 V_A}{s^2 C_1 C_2 + sC_1 gm_2 + gm_1 gm_2} \tag{3.16}$$

$$\omega_O = \frac{gm}{\sqrt{C_1 C_2}} \tag{3.17}$$

An important factor to define the selectivity of the filter is the Quality factor (Q) that is described as the resonant frequency (f_O) divided by the bandwidth (f–f_2) [73]. The higher the Q, the higher the selectivity resulting in a bandwidth decrement of the filter. In that specific design, looking for low cut-off frequencies, high values of selectivity could be achieved following Eq. (3.15).

$$Q = \frac{f_O}{f_2 - f_1} = \frac{f_O}{\Delta f} = \sqrt{\frac{C_2}{C_1}} \tag{3.18}$$

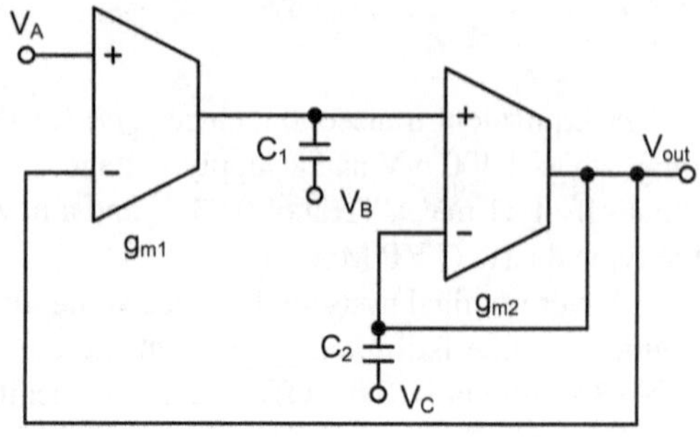

Fig. 3.25 2nd order gm-filter based on OTA amplifiers

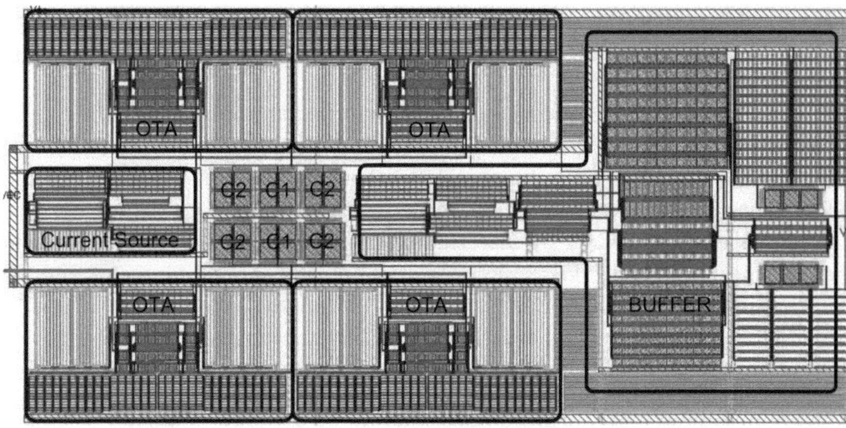

Fig. 3.26 Layout of the 4th order filter based on OTA amplifiers

For a quality factor Q of 0.7 is derived that $C_2 = 2\ C_1$ because in that design $gm_1 = gm_2 = gm$ [73]. For a cut-off frequency of 3 Hz, and $gm = 4.25$ nS, then $C_1 = 100$ pF and $C_2 = 200$ pF. These are huge values to be integrated. Then, for a test approach, $C_1 = 10$ pF has been adopted defining a cut-off frequency of 47 Hz.

Moreover, a four-order filter is designed connecting in cascade two 2nd order filters. It presents a better frequency response and roll-off decay, but in terms of area and power dissipation, it has been considered to define a cut-off frequency around 30 Hz and a $C_1 = 10$ pF. This filter has a power consumption of 50 μW, and a size of 770 μm × 575 μm. Figure 3.26 shows the layout of the 4th order LP filter. The second order filter follows the same structure with just two integrated OTA's. Finally Table 3.10 summarizes the experimental electrical features for both filters.

Table 3.10 2nd and 4th order filter characteristics

	Parameter	Value				
Filter	Type of filter	2nd order		4th order		
V_{CC}	Power voltage	1.2 (±0.6)	2.5 (±1.2)	1.2 (±0.6)	2.5 (±1.2)	V
V_{in}	Input voltage	0.8 (0.4)	1.9 (1)	0.8 (0.4)	1.9 (1)	V
F_{3db}	Cut-off freq.	47.4	49.34	31.95	23.18	Hz
g_m		4.241	4.487	4.241	4.487	nS
BP	Attenuation	>40	>60	>80	>80	dB
P_{CC}	Consumption	1.12	1.9	2.3	3.96	μW
Area	L × W (μm)	400	560	775	570	

3.5 Biotelemetry for Implanted Devices

Biotelemetry [2, 74–76] consists of transmitting the information obtained from the patient to the signal processing or reception unit using a wireless link. This technique is used in some medical instruments, like in ECG [77], in order to monitor a patient while giving full mobility to the subject. Moreover, this is one of the best methods to isolate the patient from the power lines avoiding the risk of electric shocks.

The transmission of the information is an important challenge to solve for those implanted devices under the human skin and, Biotelemetry, is one of the most suitable techniques to establish a communication between the internal detection unit, under the human skin, and the external reader.

Solutions based on wires are also studied in the bibliography but they present some technical problems, like possible infections and the inconvenience of wearing wires through the skin, and nowadays are not widely used.

Two types of Implantable devices can be distinguished: the first ones known as monitor devices [78] and the second ones known as detection devices [79].

Monitor consists on sending continuously information about the state of the patient or about the substance to be analyzed to the external reader. On the other hand, detection devices just send information about specific important points of the state of the subject or substance.

I.e. in a monitoring device the glucose is sensed continuously and the results are transmitted also continuously or periodically [80, 81]. If a detection device is used, the glucose is sensed continuously but just some important threshold values are transmitted. So, two types of data transmission modes can be defined depending on the type of the implantable device:

- Mode RAW/CONTINUOUS data transmission:

It is used in Monitor devices because it allows high data rates, lots of bidirectional information can be transmitted continuously. Normally, it uses an extra dedicated antenna for communication because the transmission frequency is higher than the used for inductive powering (Fig. 3.27).

It implements complex analog and digital modulation mechanisms [82] in order to send as much information as possible. Furthermore, time and frequency multiplexing of several signals is also possible [83]. The main disadvantage regards the high power levels required for the transmission and processing. Specific low power circuits like receivers, emitters, transponders [84] . . . have to be developed specifically for implantable devices [85, 86] to avoid problems regarding thermal transference with the biological tissue around it.

- Mode SINGLE data transmission:

It consists in sending small amount of information, just some bits or bytes. The idea is to transmit the necessary data to advice the external unit about some specific point (or points) with medical interest.

Since the transmitted data rate is low it is possible to use as a carrier frequency the same as the one used for the inductive powering [87] then, a modulation method called backscattering could be used.

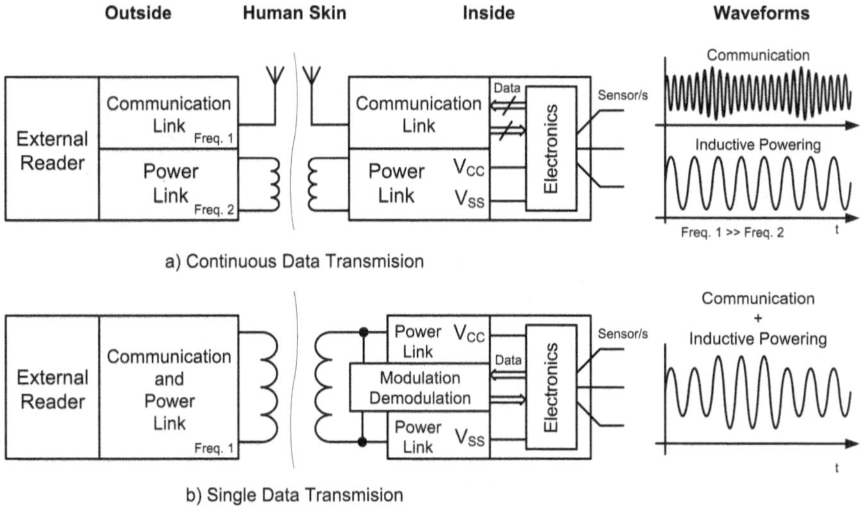

Fig. 3.27 Continuous and single communication data methods block diagram

It consists in modifying the inductive coupling between the primary and remote coil. The transmission could be done in both directions, from the primary to the secondary coil by modulating the magnetic field, and in the opposite direction changing the impedance, tuning frequency or power consumption of the secondary coil. In that way, a variation in the induced secondary coil magnetic field produces a variation of the primary coil voltage defining an AM modulation as seen in Fig. 2.27.

The circuits involved on this kind of transmission are less complex than the ones used in the continuous mode, therefore the involved power is also smaller. This is an important reason to select that mode to develop detection implantable devices. Moreover, the simple procedure for data communication using the same magnetic field created by the power transmission coils is an extra added value for implantable devices in order to reduce the total size of the device and to reduce the impact in its biological environment.

3.5.1 Low-Power Single Data Transmission Protocol

As it is introduced a Single Data Transmission consists in just sending some specific points of the evaluated data. The implemented transmission approach uses a comparator block able to detect three different threshold points and a monostable [88] based circuit to transmit the data.

The block diagram of the communication module is presented in Fig. 3.28. It is formed by 5 sub circuits, a threshold generator circuit (Vth Gen.), a low-pass filter (LPF), a Comparator Module, a DC Modulation and a monostable based circuit. The low pass filter (LPF) is used to reduce the noise in the input voltage. Then, the

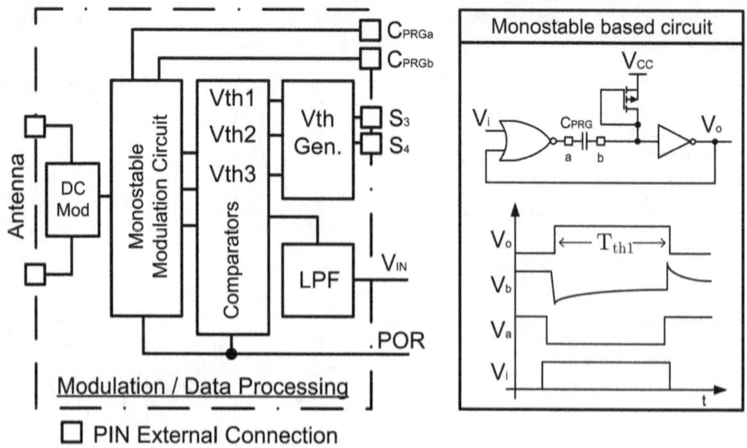

Fig. 3.28 Modulation/data processing block diagram

filtered signal is introduced to the comparator circuit where the threshold points are detected (interesting points).

Later on, and depending on the result of the comparison, the Modulation Circuit activates the modulation through the DC switch. The internal threshold voltages are generated on-chip and could be selected using signals S_3 and S_4.

The designed LPF is based on the 2nd order OTA filter presented in Section 3.4.3 with a cut off frequency of 50 Hz. The filter reduces problems regarding glitches and fast transient noise signals avoiding problems in the level detection. It reduces the risk of transmitting false positives to the main station.

The AM communication is done following the idea presented in Fig. 3.29. The circuit monitors the incoming voltage V_{IN} and, in function of three threshold voltages (V_{th1}, V_{th2} and V_{th3}), modulates the amplitude of the signal at the internal antenna.

A Power-On-Reset (POR) signal activates the circuitry when there is enough voltage to drive properly the circuits, start-up time (T_{SUP}), and the antenna starts to transmit continuously a series of ones. At that point, the Modulation/Data Processing unit is prepared to transmit the detected threshold voltages.

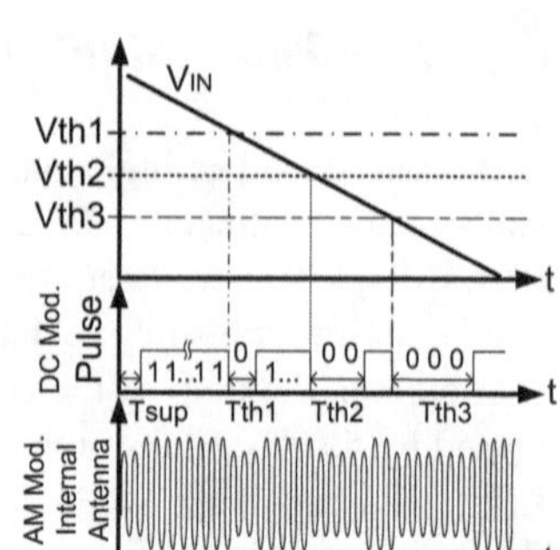

Fig. 3.29 Implemented single data transmission protocol

When the input voltage reaches the first threshold level the system transmits one zero (T_{th1}) to the external reader. Two zeros are transmitted (T_{th2}) if the second comparator threshold value is reached by V_{IN}. Finally, when the third is achieved a series of three zeros are sent (T_{th3}). Several on-chip threshold voltages could be selected using S_3 and S_4 (Section 4.2.7).

A default zero time slot interval is defined as 250 ms ($T_{th1} = 250$ ms). This time is controlled by an external capacitor, C_{PRG}, connected in the monostable based circuit. The time slot is controlled by the charging time of the C_{PRG} defined as

$$C_{PRG} \cdot \Delta V = I \cdot \Delta T \tag{3.19}$$

where I is the current injected to the capacitor by the diode connected transistor (Monostable circuit, Fig. 3.28), ΔV is the defined threshold voltage of 0.8 V used to activate the inverter gate of the monostable circuit and ΔT is the desired time slot used to define a transmitted zero. The time slot (T_{th1}) could be easily modified by changing the value of the capacitor, C_{PRG}. Table 3.11 presents some examples of capacitors and its defined time slot.

Table 3.11 Time slot (T_{th1}) definition using C_{PRG}

I(nA)	ΔV(V)	T_{th1}(ms)	C_{PRG}
1	0.8	125	150 nF
1	0.8	250	320 nF
1	0.8	500	625 nF
1	0.8	750	950 nF
1	0.8	1,000	1.25 μF

The comparisons of the input signal with the threshold voltages are achieved using three low-power comparators. The selected ones are based on the circuit explained in Section 2.3.3.2 (Fig. 2.45) with a reduced hysteresis of 5 mV. The introduction of the hysteresis helps the LPF filter to reduce noise problems with the input signal.

Then, the signals generated by those comparators are used by the logic circuitry to transmit the number of zeros. If more than one zero has to be transmitted, the circuitry re-starts again the monostable circuit. The total power consumption of this block is 1.2 μW. An example of the communication protocol mechanism is presented in Section 4.3.

3.6 Chapter Conclusions

A three electrodes amperometric sensor is formed by the Counter, the *Reference* and the *Working* electrode. It generates a current proportional to the electrochemical reaction that occurs in the W electrode. Different electrochemical techniques, Cyclic

Voltammetry, EIS... can be applied to know the behavior of the reaction for several conditions.

The Potentiostat Amplifier (PA) is the electronic circuit used to deal with the three electrode Biosensors. It is basically formed by two blocks: (1) the Control and (2) the Current Detection Module.

The Control Module regulates the voltage across the *Reference* and *Working* electrodes in order to control perfectly its voltage. This is important for the electrochemical reaction that takes place in the W electrode. Three class AB operational amplifiers have been used to implement this block. On the other hand, the Detection Module is in charge to detect the current generated by the reaction at the *Working* electrode. In this work a transimpedance amplifier is the selected option.

Two low-power potentiostats have been finally integrated on silicon. The first one is designed to work up to 2.5 V or ± 1.2 V, whereas the second one is designed for low-voltage applications with a voltage up to 1.2 V or ± 0.6 V.

Both designs are tested with a commercial three electrodes sensor and with a real electro-active substance, $K_4[Fe(CN)]_6$. The results of several Cyclic Voltammetries and amperograms are compared with the ones obtained with a commercial potentiostat. Thus, it can be assured a perfect operation of the integrated PA in the desired current ranges assuring a lower power consumption and area than the commercial one.

This chapter also introduces the design and conception of a complex impedance detection circuit called Lock-In amplifier. This circuit, theoretically introduced, is based on the use of two Synchronous Demodulated Channels to obtain the real and imaginary parts of the complex impedance. Then, the circuit is validated through several simulations.

Finally, the concept of Biotelemetry is explained and a simple low-power protocol for transcutaneous transmissions is developed. It is based on the backscattering concept that consists in changing the impedance of the implanted coil to modify the coupling factor of the inductive link.

References

1. P.H. King, R.C. Fries, *Design of Biomedical Devices and Systems*, 2nd edn. (CRC Press, Florida, USA, 2009), ISBN: 978-1-4200-6179-6
2. J.G. Webster, Medical Instrumentation Application and Design, 3rd edn. (Wiley, New York, USA, 1998), ISBN: 0-471-15368-0
3. R.P. Areny, *Sensores y Acondicionadores de Señal*. 3ª Edición, (Marcombo, Barcelona, Spain, 1998), ISBN 84-267-1171-5
4. C. Wen-Yaw, P. Arnold, W. Ying-Hsiang, T. Tseng, *A 600 μW Readout Circuit with Potentiostat for Amperometric Chemical Sensors and Glucose Meter Applications*, IEEE Conference on Electron Devices and Solid-State Circuits, EDSSC 2007, 20–22 (2007)
5. F.Z. Padmadinata, J.J. Veerhoek, G.J.A. van Dijk, J.H. Huijsing, Microelectronic skin electrode. Sens. Actuators B Chem. **1**(1–6), 491–494 (1990)
6. L. Ramasamy, *ASIC System Development of MEMS Bio-chip Analyzer with Calibration, Signal Capture and Display Circuit*, (University of Cincinnati, 2005) Available at: http://etd.ohiolink.edu/view.cgi?acc_num=ucin1127337845

7. S.K. Kailasa, S.H. Kang, Microchip-based capillary electrophoresis for DNA analysis in modern biotechnology: A review, Taylor. Francis. Sep. Purif. Rev. **38**, 242–288 (2009)
8. V.M. Ivama, S.H.P. Serrano, Rhodium – Prussian Blue modified carbon paste electrode (Rh – PBMCPE) for amperometric detection of0020hydrogen peroxide. J. Brazilin. Chem. Soc. **14**(4), (Aug 2003). ISSN: 0103–5053
9. X. Ji, C.E. Banks, A. Crossley, R.G. Compton, Oxygenated edge plane sites slow the electron transfer of the ferro-/ferricyanide redox couple at graphite electrodes. Chem. Phys. Chem. **7**, 1337–1344 (2006)
10. F. Heer et al., CMOS microelectrode array for the monitoring of electrogenic cells. Biosens. Bioelecron **20**, 358–366 (2004)
11. S.M. Radke, E.C. Alocilja, A microfabricated biosensor for detecting foodborne bioterrorism agents. IEEE Sens. J. **5**(4), 744–750 (2005)
12. D.L. McCulloch, G.B. Boemel, M.S. Borchert, Comparison of contact lens, foil, fiber, and skin electrodes for patterns electroretinograms. Doc. Ophtalmol **94**, 4 (1997)
13. O. Chailapakul, J. Promnil, M. Somasundrum, M. Tanticharoen, Immobilized K3Fe(CN)6 and glucose oxidase in polypyrrole on a gold micro-electrode and the it application as a glucose sensor. J. Sci. Res. Chula. Unit. **25**, 1 (2006)
14. S.V. Dzyadevych et al., Electrochem. Enzyme. Biosens. (2006). ISBN: 966-02-4200-X
15. F. Mizutani, E. Yamanaka, Y. Tanabe, K. Tsuda, An enzyme electrode for L-lactate with chemically amplified electrode. Anal. Chem. Acta. **117**, 153–166 (1985)
16. P.N. Bartlett, R.G. Whitaker, Strategies for the development of amperometric enzyme electrodes. Biosensors **3**, 359–379 (1987)
17. L.E. Morrison, Time resolved detection of energy transfer: Theory and application to immunoassays. Anal. Biochem. **174**, 101–120 (1988)
18. H.A. Lee, M.R.A. Morgan, Food immunoassay: Application of polyclonal, monoclonal and remobinant antibodies. Trends Food Sci. Technol. **3**, 129–134 (1993)
19. C. Dumschat et al., Pesticide-sensitive ISFET based on enzyme inhibition. Anal. Chim. Acta. **252**, 7–9 (1991)
20. P. Bergveld, Thirty years of ISFETOLOGY. What happened in the past 30 years and what may happen in the next 30 years. Sens. Actuators B **88**, 1–20 (2003)
21. E. Lorenzo et al., Analytical strategies for amperometric biosensors based on chemically modified electrodes. Biosens. Bioelectron **13**, 319–332 (1998)
22. D.B. Kell, C.L. Dave, *Conductimetric and Impediometric Devices in Biosensors. A Practical Approach*, (IRL Press, Oxford, 1990)
23. D.C. Cullen et al., Multi-analyte miniature conductance biosensor. Anal. Chim. Acta. **231**, 33–40 (1990)
24. J. Colomer-Farrarons, P. Miribel-Català, A. Saiz-Vela, J. Samitier, in *Proceeding of the 16th IEEE International Conference on Very Large Scale Integration VLSI – SOC*, A 50 μW low-voltage CMOS Biopotentiostat for low frequency Capacitive Biosensor, 2008
25. A. Gore, S. Chakrabartty, S. Pal, E. Alocilja, A multi-channel femtoampere-sensitivity conductometric array for biosensing applications. IEEE Transactions on Circuits and Systems I: Regular Papers, **53**(11), 2357–2363 (2006)
26. L.Y. Woo, L.P. Martin, R. Glass, R.J. Gorte. Impedance characterization of a model Au/Yttria-Stabilized Zirconia/Au electrochemical cell in variying oxygen and NOx concentrations. J. Electrochem. Soc. **154**(4), 129–135 (2007)
27. J.M. Flores, R.D. Romero, J.G. Llongueras, Espectroscopía de impedancia electroquímica en corrosión, Instituto Mexicano del Petróleo, UNAM, Available at: depa.pquim.unam.mx/labcorr/libro/Manual-EIS-IMP-UNAM.PDF
28. A. Lasia. *Electrochemical Impedance Spectroscopy and Its Applications. Modern Aspects of Electrochemistry*, vol. 32. (Kluwer Academic/Plenum Publisher, New York, USA, 1999), Chapter 2, p. 143

29. R.J. Reay, S.P. Kounaves, G.T.A. Kovacs, An integrated CMOS potentiostat for miniaturized electroanalytical instrumentation, IEEE International 41st ISSCC Solid-State Circuits Conference, pp. 162–163 (1994)
30. C. Berggren, B. Bjarnason, G. Johansson, Capacitive biosensors, Electroanalysis **13**(3), 173–180 (2001)
31. S. Grimnes, O.G. *Martinsen, Bioimpedance and Bioelectricity Basics*, 2nd edn. (Academic Press, Elseiver, London, UK, 2008). ISBN: 0-12-303260-1
32. Diffusion Element, http://www.consultrsr.com/resources/eis/diffusion.htm
33. M.N. Latto, *The Electrochemistery of Diamon*, (University of Bristol, Bristol, UK, Sep 2001), http://www.chm.bris.ac.uk/pt/diamond/mattthesis/contents.htm
34. C.G. Zoski, *Handbook of Electrochemistery*, (Elsevier, The Netherlands, 2007), ISBN: 0-444-51958-0
35. Cypress Systems Cyclic Voltammetry, http://www.cypresssystems.com/Experiments/cv.html Amperometri
36. GAMRY Instruments App. Note, Electrochemical Impedance Spectroscopy, http://www.gamry.com/App_Notes/EIS_Primer/EIS_Primer.htm
37. J. Braz, Rhodium–prussian blue modified carbon paste electrode (Rh-PBMCPE) for amperometric detection of hydrogen peroxide. J. Brazilian Chem. Soc. **14**, 4 (2003). ISSN 0103–5053
38. A.J. Bard, L.R. Faulkner, *Electrochemical Methods. Fundamentals and Applications*, 2nd edn. (Wiley, New York, NY, 2001). ISBN 0-471-04372-9
39. S.M. Martin, F.H. Gebara, T.D. Strong, R.B. Brown, A low.voltage, chemical sensor interface for system-on-chip: The fully-differential potentiostat, Proceeding of the 2004 International Symposium on Circuits and Systeems ISCAS'02, vol. 4, pp. IV–892–895, (2004)
40. E. Lauwers, J. Suls, W. Gumbrecht, D. Maes, G. Gielen, W. Sansen, A CMOS multiparameter biochemical microsensor with temperature control and signal interfacing. IEEE J. Solid-State Circuits **36**, 12 (2001)
41. S.M.R. Hasan, Stability analysis and novel compensation of a CMOS current-feedback potentiostat circuit for electrochemical sensors. Sens. J. IEEE. **7**(5), 814–824 (May 2007)
42. J. Colomer-Farrarons, P. Miribel-Català, A. Saiz-Vela, I. Rodriguez, J. Samitier, in *Proceedings of the IEEE MWSCAS Conference*, A low power CMOS Biopotentiostat in a Low-Voltage 0.13 μm Digital technology, Cancún, Mexico, 2009
43. J. Colomer-Farrarons, P. Miribel-Català, I. Rodriguez, J. Samitier, in *Proceedings of the XX IECON Conference*, CMOS Front-end Architecture for In-Vivo Biomedical Implantable devices, Porto, Portugal, 2009
44. J. Colomer-Farrarons, P. Miribel-Català, A. Saiz-Vela, M. Puig, J. Samitier, A. Errachid, A 50 μW low-voltage CMOS Biopotentiostat for low-frequenciy Capacitive Biosensor, Proceedings of the XX IEEE VLSI Conference, Rhodes, Greece, 2008
45. S.M. Martin, F.H. Gebara, B.J. Larivee, R.B. Brown, A CMOS-integrated microinstrument for trace detection of heavy metals. IEEE J. Solid-State Circuits **40**(12), 2777–2786 (2005)
46. T.D. Strong, S.M. Martin, R.F. Franklin, R.B. Brown, in *Proceedings of the IEEE International Symposium on Circuits and Systems*, Integrated electrochemical neurosensors, 2006, pp. 4110–4113
47. R. Jacob Baker, *CMOS: Circuit Design, Layout, and Simulation*, Revised 2nd edn. (Wiley – Interscience, NJ, USA, 2008). ISBN 978–0-470-22941-5
48. R. Gregorian, *Introduction to CMOS Op-Amps and Comparators*, (Wiley, New York, USA, 1999). ISBN 0-471-31778-0
49. J.H. Huijsing, *Operational Amplifiers, Theory and Design*, (Kluwer Academic Publishers, Dordrecht, The Netherlands, 2001). ISBN: 0-7923-7284-0
50. F. Maloberti, *Analog Design for CMOS VLSI Systems*, (Kluwer Academic Publishers, The Netherlands, 2001). ISBN: 0-7923-7550-5

51. A.C. Patil, F. Xiao, M. Mehregany, S.L. Garverick, Fully-monolithic, 600°C differential amplifier in 6H-SiC JFET IC technology, Custom Integrated Circuits Conference, CICC'09, pp. 73–76, (2009)
52. J.P. Close, F. Santos, in *Proceeding of the Bipolar/BiCMOS Circuits and Technology Meeting.* A JFET input single supply operational amplifier with rail-to-rail output, pp. 149–152, (1993)
53. F. Serra-Graells, A. Rueda, J.L. Huertas, *Low-Voltage CMOS Log Companding Analog Design*, (Springer, Netherlands, 2003). ISBN: 978-1-4020-7445-5
54. R.J. Baker, *CMOS: Mixed-Signal Circuit Design*, 2nd edn. (Wiley – IEEE Press Series on Microelectronic Systems, NJ, USA, 2008). ISBN: 978-0-470-29026-2
55. National Instruments, Lab View software, http://www.ni.com/labview/
56. X. Gan, Y. Wu, L. Liu, W. Hu, Effects of $K_4Fe(CN)_6$ on electroless copper plating using hypophospite as reducing agent. J. Appl. Electrochem. **37**, 899–904 (Springer, Apr 2007)
57. CH Instruments, http://www.chinstruments.com/
58. BVT Technologies, http://www.bvt.cz/
59. N.I. Bojorge Ramírez, M.Fortes, A.M. Salgado, B. Valdman, Construction of an Amperometric Immunosensor Using Solanum Tuberosum Potato Apyrase for the Detection of Schistosomiasis. Informaci6n Tecnol6gica **20**, 3 (2009). ISSN: 0718–0764
60. K.K. Kasem, S. Jones, Platinum as a reference electrode in electrochemical measurements, Platinum Metal Rev. **52**, 100–106 (Apr 2008)
61. H.E.A. Ferreira, D. Daniel, M. Bertotti, E.M. Richter, A novel disposable electrochemical microcell construction and characterization, J. Brazilian Chem. Soc. **19**, 8 (2008)
62. I. Bontidean, C. Berggren, G. Johansson, E. Csöregi, B. Mattiasson, J.R. Lloyd, K.J. Jakeman, N.L. Brown, Detection of heavy metal ions at femtomolar levels using protein-based biosensors. Anal. Chem. **70**(19), 4162–4169 (1998)
63. E. Katz, I. Willner, Probing biomolecular interactions at conductive and semiconductive surfaces by impedance spectroscopy:Routes to impedimetric immunosensors, DNA-sensors, and enzyme biosensors, Electroanalysis **15**(11), 913–947 (2003)
64. L. Yang, Y. Li, C.L. Griffis, M.G. Johnson, Interdigitated microelectrode (IME) impedance sensor for the detection of ciable Salmonella typhimurium, Biosens. Bioelectron **19**, (10), 1139–1147 (2004)
65. B. Robert, C. Northrop, in *Analysis and Application of Analog Electronic Circuits to Biomedical Instrumentation*, ed. by M.R. Neuman. The Biomedical Engineering Series (CRC Press, Florida, USA, 2004)
66. A. De Marcellis, G. Ferri, M. Patrizi, V. Stornelli, A.D' Amico, C. Di Natale, E. Martinelli, A. Alimelli, R. Paolesse, An integrated analog lock-in amplifier for low-voltage low-frequency sensor interface, in *Proceddings of the Interational Workshop on Advances in Sensors and Interface, IWASI*, pp. 1–5 June 2007
67. D. Rairigh, A. Mason, C. Yang, Analysis of on-chip impedance spectroscopy methodologies for sensor arrays. Sensor. Lett. **4**(4), 398–402 (2006)
68. A.E. Moe, S.R. Marx, I. Bhinderwala, D.M. Wilson, A miniaturuzed lock-in amplifier design suitable for impedance measurements in cells. Proc. IEEE Sens. **1**(24–27), 215–218 (AUTRICHE 2004)
69. W. Xu, E.G. Friedman, Clock Feedthrough in CMOS analog transmission gate switches. Anal. Int. Circuits and Signal Process **44**, 271–281 (2005)
70. R. Hogervost, J.P. Tero, R.G.H. Eschauzier, J.H. Huijsin. A compact power-efficient 3 V CMOS rail-to-rail input/ouput omperational amplifier for VLSI cell libraries. IEEE J. Solid-State Circuits **29**, 1505–1513 (1994)
71. A. Veeravalli, E. Sánchez-Sinencio. J. Silva-Martínez, Transconductance amplifiers with very small transconductances: A comparative design approach. IEEE J. Solid-State Circuits **37**(6), 770–775 (June 2002)
72. A. Arnaud, R. Fiorelli, C. Galup-Montoro, Nanowatt, sub-nS OTAs, with Sub-10-mV input offset, using series-parallel current mirrors. IEEE J. Solid-State Circuits **41**(9), 2009–2018 (Sept 2006)

73. J.M. Fiore, in *Amplificadores Operacionales y Circuitos Integrados Lineales*, ed. by Thomson. (Mohawk Valley Community College, Ed. Paraninfo, Madrid, Spain, 2002)
74. R.S. Machay, *Bio–Medical Telemetry*, 2nd edn. (Wiley, New York, NY, 1970)
75. H.P. Kimmich, in Biotelemetry, eds. by J.G. Webster. *Encyclopedia of Medical Devices and Instrumentation*, (Wiley, New York, NY, 1980)
76. A. Santic, M.R. Neuman, *A low-power infrared biotelemetry system, Biotelemetry VIII*, (Kimmich/Klewe, Netherlands, 1984)
77. S.A.P. Haddad, W.A. Serdijn, Ultra low-power biomedical signal processing: An analog wavelet filter approach for pacemakers, Anal. Circuits and Signal Process. (Springer 2009). ISBN: 978-1-4020-9072-1
78. M.R. Haider, S.K. Islam, M. Zhang, A low-power signal processing unit for in vivo monioring and transmission of sensor signals. Sens. Trans. J. **84**(10), 1625–1632 (2007)
79. Positive ID/Verichip White Paper, Development of an Implantable Glucose Sensor, http://www.positiveidcorp.com/white-papers.html
80. K. Van Schuylengergh, R. Puers, *Inductive Powering. Basic Theory and Application to Biomedical Systems*, (Springer, The Netherlands, 2009). ISBN: 978-90-481-2411-4
81. B. Lenaerts, R. Puers, *Omnidirectional Inductive Powering for Biomedical Implants*, (Springer, The Netherlands, 2009). ISBN: 978-1-4020-9074-5
82. W.C. Lin, S.K. Pillay, A micropower pulsewidht-modulation-pulse-position-modulation two-channel telemetry system for biomedical applications. IEEE Trans. Biomed. Eng. **BME – 21**, 273–280 (1974)
83. C. Weller, Electrocardiography by infrared telemetry. J. Physiol. (London) **267**, 11–12 (1977)
84. Z. Tang, B. Smith, J.H. Schild, P.H. Peckham, Data transmission from an implantable biotelemeter by load-shift keying using circuit configuration modulator. IEEE Trans. Biomed. Eng. **BME-42**, 524–528 (1995)
85. A. Santic, M.R. Neuman, *A low-power infrared biotelemetry system, Biotelemetry VIII*, (Kimmich/Klewe, Netherlands, 1984)
86. H.P. Kimmich, *Biotelemetry, Enciclopedia of Medical Devices and Instrumentation*, (Willey, New York, USA, 1988), pp. 409–425
87. N. Donaldson, Passive signalling via inductive coupling. Med. Biol. Eng. Comput. **24**, 223–224 (1986)
88. D. Gajski, *Principios de Diseño Digital*, (Prentice Hall, Madrid, España, 2000). ISBN: 84-8322-004-0

Chapter 4
CMOS Front-End Architecture for In-vivo Biomedical Subcutaneous Detection Devices

Abstract This chapter describes the design and conception of the Self-Powered CMOS Front-End Architecture for a Biomedical Subcutaneous Device. The entire architecture is presented in detail as well as the powering and communication through the inductive link. The power and communication antenna and the connections between the MHCP IC (Chapter 2), the BioChip IC (Chapter 3) and the sensor are also detailed afterwards. The results obtained with the final capsule prototype with a size less than 4.5 cm $\times$ 2.5 cm are shown and commented in depth. Problems regarding misalignments between the internal and external antennas are studied and the SOA (Safety Operation Area) region is introduced. Finally, the prototype has been validated as a detector.

Keywords Low-power instrumentation amplifier · Wireless implanted devices · Self-powered device · Event detector implantable device · Inductive power transmission · Biomedical telemetry

4.1 Introduction

The integration of medicine and electronic technologies allows the development of biomedical devices able to diagnose and/or treat pathologies by detecting and/or monitoring pathogens, multiple ions, PH changes, and so on. Nowadays, the advances in different areas such microelectronics, microfluidics, microsensors and biocompatible materials open the door to develop human body Lab-on-Chip implantable devices and Point-of-Care in vitro devices [1, 2, 3]. As a result there are continuous monitoring systems able to develop clinical tasks in less time and for lower-cost that the usual tests, like ELISA technique [4].

Different problems should be overcome to obtain the ideal implantable device [5]. The device must be biocompatible to avoid undesirable reactions within the body. Furthermore, it must provide a long-term stability, selectivity, calibration, miniaturization, and reproducibility and power in a downscaled and portable device.

Power and communications are also key aspects in the conception of an implantable device. The first one resides in the way to transfer enough energy to power the device, whereas the second one consists in the integration of the necessary

J. Colomer-Farrarons, P.L. Miribel-Català, *A CMOS Self-Powered Front-End Architecture for Subcutaneous Event-Detector Devices*, DOI 10.1007/978-94-007-0686-6_4,

instrumentation and communication electronics to control the sensors and to send the information provided by the sensors through human skin.

An inductive coupling RF power harvesting is a growing alternative method for transmitting energy to the implanted device, substituting the use of batteries or wires [4, 6, 7]. Furthermore, this alternative permits to establish a bidirectional communication between the implanted device and an external interface (base or reader) [8–12].

The basics of this RF mechanism are the same as the RFID tags, where the implanted device acts as a tag. Some implantable telemetry circuits based on inductively coupling can be found in the literature like in [13–16].

Although, a lot of works are focused on the development of robust electronics for in-vivo monitoring, this means, continuously sensing, recording and data transmission of the obtained information of the studied substance, not all the in-vivo applications need these electronics. There are other less studied applications more focused on the detection of presence or absence of a certain levels of proteins, antibodies, ions, oxygen, glucose, etc.

These in-vivo event detection circuits, or True/False applications [16], work as an alarm; when the analyzed concentration level exceeds, under or over, a threshold value or the system detects the presence of zero-tolerance pathogens like salmonella [16], the system sends to an external reader a signal indicating the fault or showing the value as in some pregnancy tests. 230

For instance, in the case of glucose monitoring, such as the detection of a threshold decrease in the glucose level, it is mandatory to avoid critic situations like the Hypoglycemia [11, 12, 17]. There are different approaches for continuous glucose monitoring. From commercial solutions, like the blood glucose tester by Cygnus Inc. [18], to subcutaneous Minimed Medtronic [18] and Abbott Inc. [19] solutions, which control the glucose level every 3–5 min. They are placed just under the skin with a closed-loop control to delivery insulin and an autonomy of 3–5 days because biofouling problems. Otherwise, solutions looking for a minimum biological impact are also reported [20–22] including an inhibitor to resist biofouling, coating needle type electrochemical sensors.

In terms of the sensors, electrical Biosensors like label-free Biosensors are great candidates to be integrated due to their low cost, low power and easy miniaturization. Recent developments applied in the field of glucose monitoring [23, 24], pregnancy test or DNA [25] are examples of the potentiality of these electrical nanobiosensors. Their electrical behavior could be voltammetric, amperometric or an impedance variation.

In some sense and depending on the type of application, it is not necessary to measure and send raw (continuous) data with high accuracy, from the internal mote to the external unit, when the detection of vital signs or threshold values is enough for the monitoring purpose. Hence, implantable detection devices working as alarm or on/off detectors with some local processing would save power and communication requirements.

The implantable system in Fig. 4.1 will be a platform to perform a true/false alarm monitoring for different targets, which will be transferred to a central data-base where all the inputs will be personalized for each patient. The collected

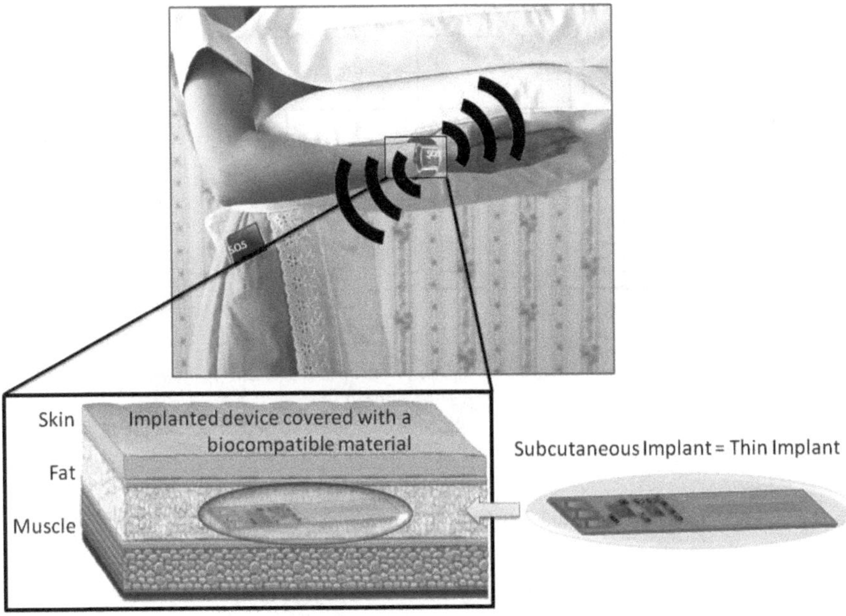

Fig. 4.1 Conception of the implantable detection device

data will be measured at different scenarios: when the patient is at rest, doing some type of physical activity, which presents some medical interest for good prognostic and diagnostic outcomes [26].

The system will be concerned in research as a complement to study constantly the patients when they are carrying out their usual daily activities in normal conditions (outdoors). In that way, it is possible to avoid secondary effects like psychological alteration caused by the stress of being in a hospital with unknown people etc...

At this stage, it is proposed a generic implantable front-end architecture based on inductive coupling for in-vivo presence or absence detection of pathogens, ions, glucose, oxygen concentration, etc, for a generic amperometric Biosensor application. The in-vivo event detection circuits [16] work as an alarm. When the analyzed concentration level is out of a range of accepted values, a threshold value activates the alarm. The detection will take place when the amplitude of the measured signal crosses a specified threshold value [27].

Two final prototypes have been developed to validate the conception of the integrated device. These prototypes combine two full custom IC's, MHPC and BioChip, the Inductive link with a 13.56 MHz antenna, and the connections for the three electrodes amperometric sensor.

4.2 Front-End General Architecture

A general structure of blocks for an implantable device powered by an inductive coupling is presented on Fig. 4.2. It is comprised of a Biosensor, an antenna and eight electronic modules. A three electrodes amperometric sensor has been used

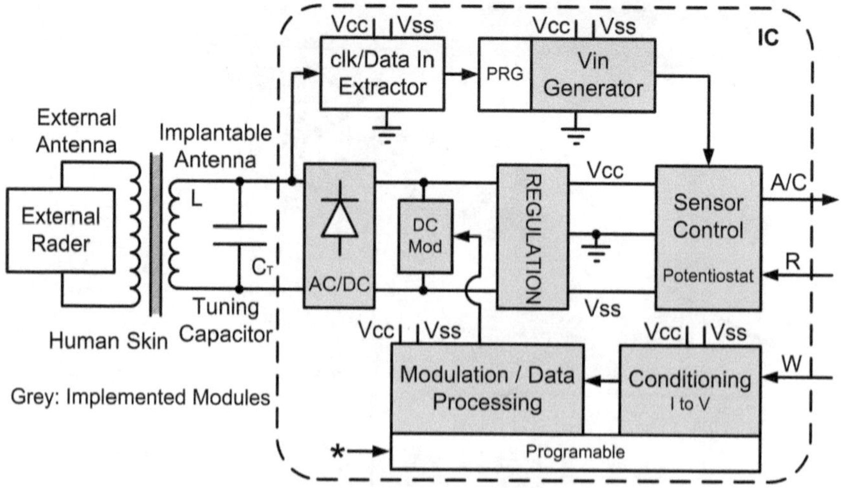

Fig. 4.2 Proposed generic implantable front-end architecture

to explain and develop the system. Anyway, it is easy to adapt the topology for any kind of sensor by simply modifying two modules: the Sensor Control and the Sensor Conditioning.

The three electrodes that compose the sensor are: (a) the *Working electrode* (W), which serves as a surface on where the electrochemical reaction takes place; (b) the *Reference electrode* (R), used to measure the potential at the W electrode, and (c) the *Auxiliary or Counter* electrode (A/C), which supplies the needed current required for the electrochemical reaction at the W electrode (Section 3.2.1).

The system is conceived as an active RFID Tag [22, 28] where the inductively coupled link, generated by the implantable and by the external antenna, is able to supply enough energy to power all the system. It provides a wireless bidirectional communication through the human skin, transmitting in this way the information obtained by the sensor and receiving data from the external reader to configure the implanted electronics [29, 30].

As a detector of absence or presence of any substance the system works as follows. When the inductive link delivers enough (plenty) energy and the on-chip voltage regulators work properly, a power-on signal activates the electronics. At this point the Sensor Control module drives the amperometric Biosensor with the voltage provided by the Vin Generator. When this voltage is applied between the R and W sensor electrodes the reaction takes place and the sensor starts to generate a current proportional to the reaction.

This current is then measured and conditioned by the Sensor Conditioning module and the signal is forwarded to the Modulation/Data Processing module. In this block, the absence/presence detection takes place as well as the modulation process to send the information to the external reader using a backscattering method.

The *Modulation/Data Processing* module has a programmable threshold values. The module activates the communication using a backscattering method through the inductive link when the current generated by the sensor achieves the threshold values. In this way, the external reader can be quickly advised every time the desired substance exceeds the programmed threshold levels. The role of each module and the selected implementation are described in the following:

4.2.1 Antenna or Induction Coil

The antenna (or inductive coil) is used as a wireless communication link and power energy source. It should produce energy to supply the electronics and should have enough communication range to send and receive data through the human skin [31, 32].

Several possibilities could be explored to design an implantable antenna. It is possible to use a small coiled antenna with extremely small size like those used in some RFID human or animal identifiers by Verichip® Corp. Also it is doable to integrate an antenna using a MEMS CMOS compatible process or using a modern organic technology like in [15].

As a first approach, the same rectangular coil as the ones used in Section 2.2.1 has been first selected to build up a PCB based setup to validate each block and to characterize the whole system. It has three turns, size of 30 mm × 15 mm, 0.5 mm thickness, a conductor width of 1 mm and an inductance of 220 nH. It is tuned to work at the RFID frequency of 13.56 MHz.

One small antenna has been designed to miniaturize the system and to validate the architecture as implantable device. The prototype is a rectangular coil of 5.5 mm × 14.5 mm and a thickness of 0.5 mm. It has seven turns, a conductor width of 0.2 mm, inductance of 400 nH and a series resistance of 340 mΩ. Since a subcutaneous implant would be designed, a rectangular thin coil could be one of the best's options as is depicted in Fig. 4.1.

On the other hand, a second wire-coiled antenna of copper with 2.6 mm of diameter has been fabricated just to compare the electrical and mechanical characteristics. It has a length of 7.5 mm, 0.071 mm copper diameter, 14 μH, and a parasitic resistance 5 Ω. Both small antennas are depicted in Fig. 4.5. Table 4.1 presents the electrical and mechanical parameter for the two miniaturized antennas.

Both antennas are tuned to work at the RFID frequency of 13.56 MHz. A minimum energy loss due to the skin absorption could be achieved working at this frequency range [33] but, there are other available frequencies that could also be used.

Table 4.1 Small implantable antennas parameters

Symbol	Parameter			Units
Type		Rectangular	Coiled	
L	Inductance value	400 n	14 μ	H
C_T	Tuning capacitor	350 p	10 p	F
R_P	Parasitic resistance	340 m	5	Ω
W_t	Width	5.5	–	mm
L_t	Length	14.5	7.5	mm
D	Diameter	–	2.6	mm

4.2.2 AC/DC and Regulation Module

The coupled antenna produces an AC voltage; hereby an AC/DC rectifier is necessary to generate an unregulated DC voltage. A regulation block produces a regulated DC voltage to drive all the electronics. Unipolar or bipolar power supply architecture could be selected. A bipolar power scheme able to supply a regulated voltage up to ±1.2 V and a maximum current of ±1.5 mA has been implemented.

The AC/DC block is based on a half-bridge NMOS rectified with a bulk control voltage introduced in Section 2.3.2.2. Then, two LDO's [34] and two Complex Control Modules (CCM – Section 2.3.3.2) are connected to the power scheme. In that way, the DC voltage is regulated and controlled. The LDO's are based on the structure presented in Section 2.2.2; just a small variation on its structure is applied for the negative one [34].

The Control Modules are used basically for two main reasons: firstly, they activate a Power-On Reset (POR) signal when minimum voltages of 0.8 V and –0.8 V are achieved. This signal is then used to activate the rest of the internal circuitry. Secondly, both Control Modules generate several Bandgap reference voltages that are used in subsequent circuits. Moreover, it is a simple method to program the voltage from ±0.6 V to ±1.2 V in order to drive properly the circuits. Figure 4.3 presents the architecture of the power stage.

Further on in this chapter, it is presented the obtained generated on chip voltages at various air coupling distances as well as other experimental measurements using the miniaturized antennas.

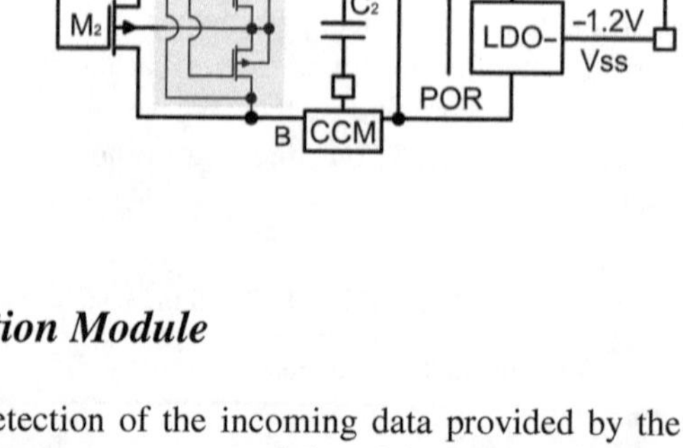

Fig. 4.3 AC/DC integrated rectifier architecture

4.2.3 Data and Clock Extraction Module

This block is responsible for the detection of the incoming data provided by the external reader and, if it is the case, programs some modules of the scheme, besides it recovers and produces the necessary clocks used by the electronic modules. Several approaches can be found in the literature [14] to design this circuitry.

This module has not been implemented in this approach because no internal clocks are necessary and no programmable capabilities are included in any block.

4.2.4 Vin Generation Module

This block is used to generate the signals to be applied to the sensor. In function of the performance of the integrated Biosensor and how it works, the block can generate a DC or AC signal. This module could be programmed by the external reader in order to add more functionality to the system.

In the presented approach, three internal signals and one external signal can be selected. The selection is done manually by two external selector pins S_1 and S_2, Fig. 4.4. Table 4.2 shows the values of the internal generated voltages. Thanks to the external connection it is possible to introduce an off-chip AC or DC signal or directly ground the signal to Gnd.

Programmability functions and digital voltage generation could be introduced in order to add more functionality to the system to fix the DC levels for the amperometric analysis. However, the use of complex generation circuits introduce higher power consumption and hence, problems regarding temperature dissipation around the biological environment could arose.

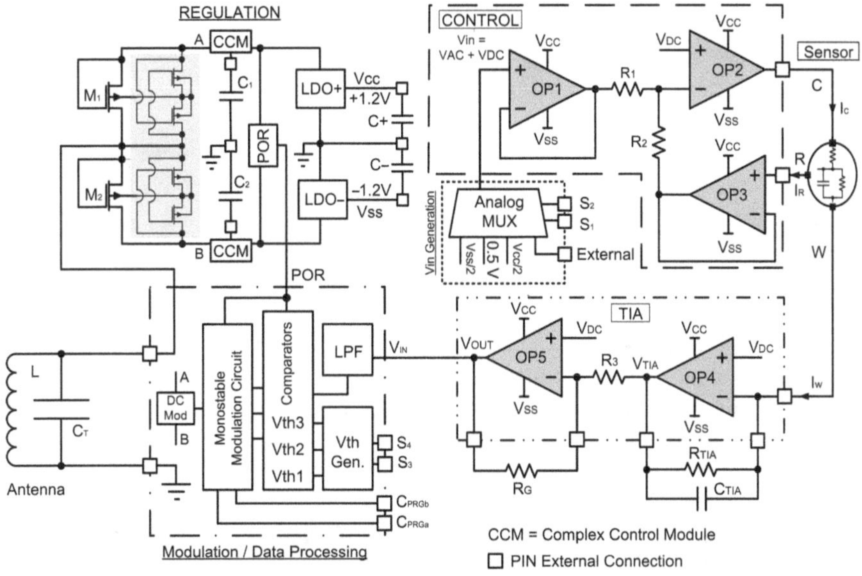

Fig. 4.4 Schematic of the implemented architecture for the in-vivo implantable detectors (V_{DC} connected to Gnd)

Table 4.2 On-chip voltage generation

S_1	S_2	Voltage
Gnd	–	$V_{CC}/2$
–	Gnd	$V_{SS}/2$
Gnd	Gnd	0.5 V
–	–	External

Three internal voltages of 0.6, –0.6, and 0.5 V can be selected in the final implementation. Also, an external voltage could be used. The Regulation Module generates the internal reference voltages.

4.2.5 Sensor Control Module

This circuitry applies the voltage provided by the Vin Generation Module –Vin- to the sensor, between the Sensor Cell Electrodes.

The topology of the circuitry should be selected in function of the used type of sensor. This work is focused to deal with three electrodes sensors, so the adopted module is based on the potentiostat control module presented in Section 3.3.1. A current through the cell is defined in such a way that the voltage difference between the *R* and *W* electrodes follows the defined V_{in} signal, which polarizes the sensitive cell.

The implemented architecture is based on the low power ±1.2 V potentiostat in Table 3.5 (Section 3.3.3). It works perfectly with DC signals and V_{in} voltages up to ±1 V. Figure 4.4 depicts the potentiostat control module and its connections.

4.2.6 Conditioning Module

This block detects the signal generated by the amperometric Biosensor and adapts it to be treated by the subsequent module. A current-to-voltage converter based on the transimpedance amplifier –TIA- in Section 3.3.2 is the selected option for this approach.

As it is detailed in the previous chapter, the current-to-voltage conversion is defined by $V_{TIA} = -I_W R_{TIA}$, where I_W is the current flowing in the working electrode and R_{TIA} is the externally selectable transimpedance resistance. A second gain stage, based on an inverter configuration, adapts the voltage values to feed the next stage. The gain is defined by R_G and is externally selectable. Figure 4.4 presents the TIA and the gain amplifier.

4.2.7 Modulation and Data Processing Module

The data generated by the sensor and adapted by the Conditioning Module is then analyzed and emitted to the external reader through this analogue or module. Depending on the application, a complex analysis should be needed using some digital processing that increases the consumption and complexity of the module [35–39].

For an integrated detector, the use of comparators is a smart solution to detect one or several points with medical interest. In that approach, comparators able to detect three different threshold voltages (Fig. 4.4, V_{th1}, V_{th2} and V_{th3}) have been designed for its simplicity and low power consumption.

Table 4.3 On-chip generated threshold voltages

S_3	S_4	V_{th1}	V_{th2}	V_{th3}
Gnd	–	–500 mV	–250 mV	–125 mV
–	Gnd	–125 mV	–250 mV	–500 mV
Gnd	Gnd	125 mV	250 mV	500 mV
–	–	500 mV	250 mV	175 mV

All three-threshold voltages, V_{th1}, V_{th2} and V_{th3}, are generated on-chip and programmed externally using S_3 and S_4. A modulation circuitry activates the backscattering communication in function of the comparators response. This method consists in changing the impedance of the antenna, so the external antenna can detect this variation.

A DC modulation [15] is used implementing a PMOS (W/L = 3,000/2 μm) transistor connected directly behind the rectifier. This modulation presents a higher Q-factor modulation than the AC [10].

The Modulation/Data Processing block is based on the circuit presented in Section 3.5.1. The communication time slot is controlled by an external capacitor, C_{PRG}, connected to the monostable-based circuit. This capacitor controls the width of the transmitted zero.

Table 4.3 shows the on-chip generated threshold voltages in function of pins S_3 and S_4 whereas Fig. 4.4 presents the connections of the block with the rest of the system.

4.2.8 External Reader

As it is previously explained, the external reader transmits the power to the integrated antenna and receives the data from it. The adopted solution is based on the use of the commercial demo board TRF7960EVM [40] by Texas Instruments®.

This evaluation module uses the TRF7960 multi-standard fully integrated 13.56 MHz RFID reader chip able to transmit with a selectable output power of 100 or 200 mW. Figure 4.7 presents the TRF7960EVM board with its PCB antenna.

4.3 Prototypes Design and Results

4.3.1 Prototypes

Two miniaturized devices have been fabricated to validate the implantable detector architecture. Both devices are presented in Fig. 4.5. The first one uses a rectangular antenna whereas the second one is designed with a coiled antenna. The parameters of both antennas are presented in Table 4.5.

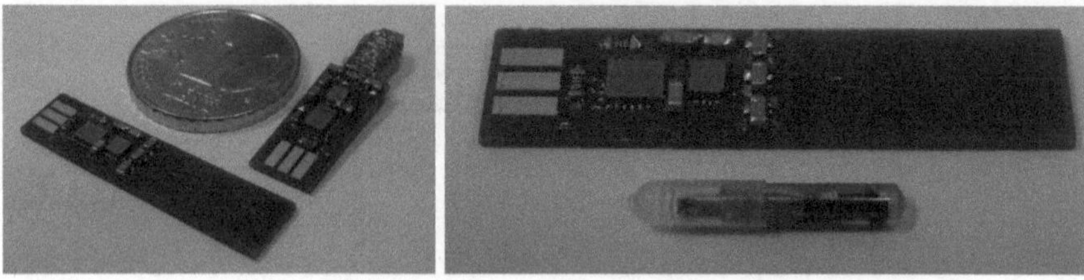

Fig. 4.5 (*Left*) fabricated prototypes (rectangular coil – bottom- and coiled antenna) and (*right*) size comparison of the rectangular prototype and a commercial RFID pet identifier by Indexel-Merial

Each prototype is mounted on a FR4 substrate PCB with a thickness of 0.5 mm. It is composed by two IC's, one for powering and communications and the other one with the sensor instrumentation and post processing, one antenna, a sensor connection and the external capacitors and resistors. The IC's are designed in 0.13 μm technology from STMicroelectronics®.

The size of each off-chip component is 1.6 mm × 0.8 mm that corresponds to the 0603 SMD passive components package. Figure 4.6 depicts the distribution of the components in the PCB.

The sensor/Biosensor connections are located on the right side of the device, Fig. 4.6, with a pad size of 1.1 mm × 3.4 mm. These pads could be used as a connection to other sensor or directly as gold electrodes.

The total size of the rectangular prototype is 5.5 mm × 29.5 mm × 1.5 mm. On the other side, the coiled prototype has an area of 5.5 mm × 23 mm × 3 mm. Finally, Table 4.4 summarizes the features of the fabricated prototypes.

It is important to note that all circuits were tested individually before manufacturing the two prototypes using a setup based on the PCB and LabVIEW® software presented in Section 3.3.3.

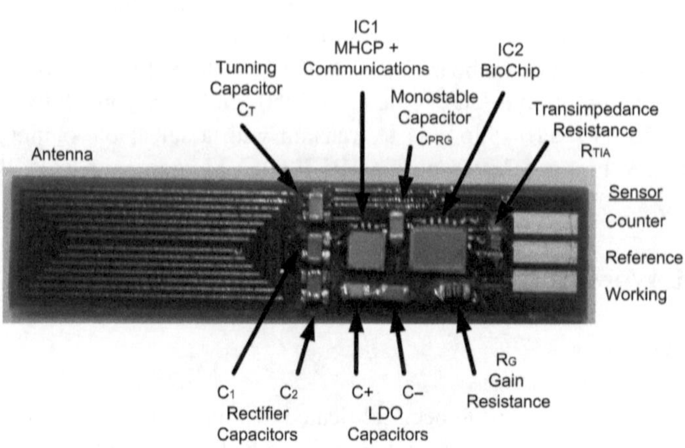

Fig. 4.6 Components placement of the fabricated prototypes. Both prototypes present the same distribution. (IC1 – powering, IC2 – sensor control and signal conditioning)

Table 4.4 Mechanical characteristics of the two fabricated prototypes

Symbol	Parameter	Value		Units
Type	Antenna type	Rectangular	Coiled	
L	Inductance value	400 n	14 μ	H
C_T	Tuning capacitor	350 p	10 p	F
C_1	Rectifier capacitor	4.7 u		F
C_2	Rectifier capacitor	4.7 u		F
C+	LDO capacitor	1 u		F
C-	LDO capacitor	1 u		F
C_{PRG}	Monostable capacitor	330 n		F
R_{TIA}	Transimpedance resistance	25 k		Ω
R_G	Gain resistance	1 M		Ω
	Sensor material	Gold		
A	Area	5.5 × 29.5 × 0.5	5.5 × 23 × 0.5	mm^3
R_I	Integrated resistances $R_1 = R_2 = R_3$	1 M		Ω

4.3.2 Experimental Results

The in-vivo detection architecture is validated using the rectangular prototype as implantable device. The Texas Instrument® TRF7960 works as external reader with a maximum emission power of 200 mW at 13.56 MHz and the commercial sensor AC1.W1.R1 by BVT Technologies® acts as a Biosensor. The ±1.2 V Potentiostat Amplifier (PA) is used to drive the Biosensor.

A wired connection is done to connect the AC1.W1.R1 sensor with the sensor pads of the implantable device. Thus, it is possible to manipulate the device under test (DUT) while the sensor always remains fixed.

First of all, the recovered voltage in the implantable device is studied in function of the distance (Z Axis) between the reader and the implantable prototype and in function of the XY misalignments. Figure 4.7 shows the origin of XY coordinates used in the study. The center of the implantable antenna (black cross) is positioned in the (0.0) coordinates of the emitter. The Z axis is orthogonal to XY.

Figure 4.8 presents the dependence of the rectified voltage in function of the Z axis, the distance between both antennas. The maximum allowable height (Z axis separation) is around 20 mm. In that point, the rectified voltage (V_{rec}) is not enough to drive properly the on-chip LDO`s thus, it is not possible to generate the regulated voltages, V_{CC} and V_{SS} (±1.2 V).

This 20 mm distance is also the maximum available distance to transmit properly information from the implantable device to the external reader. The recovered power is in the range of 1.1 mW to 800 μW from 1 to 20 mm respectively.

The next step consists on evaluating the XY misalignment. Figure 4.9 depicts the rectified voltage (V_{rec}) distribution in function of the XY misalignment for Z distances of 10, 15, and 20 mm. The voltages obtained in the (0.0) position are equal to those presented in Fig. 4.8 for all Z distances so, it can be stated that the farther the antenna is placed from the center the lower is the rectified voltage.

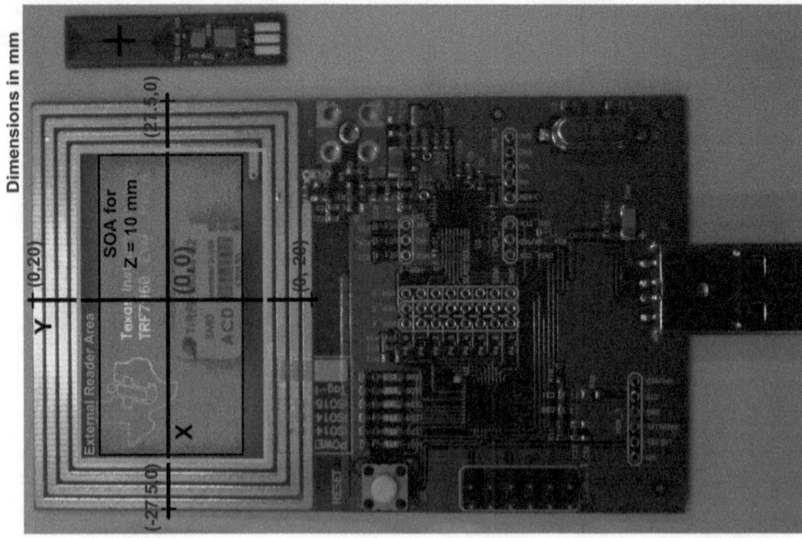

Fig. 4.7 Prototype of the implantable device (*top*, 5.5 mm × 29.5 mm) and the TRF7960 transponder used as external reader. It is depicted the XY coordinates used to evaluate the effect of the misalignment. The implanted device moves around the external reader area (ERA) for difference height distances (Z axis distance) to define the safety operation area (SOA); the area where the device can move while working properly

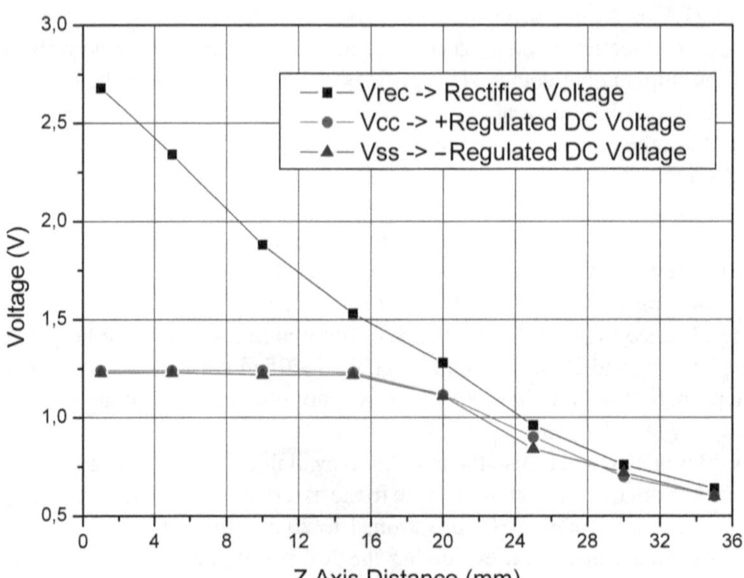

Fig. 4.8 Recovered rectified voltage (V_{rec}) in function of the Z distance between the implantable and the external antenna at (0.0) in air coupling conditions. It is also depicted the on-chip regulated voltages, V_{CC} and V_{SS}

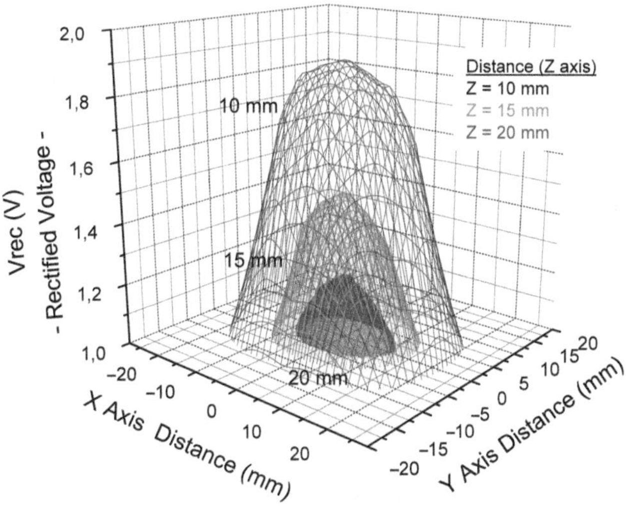

Fig. 4.9 Recovered rectified voltage distribution in function of the X-Y misalignment for Z distances of 10, 15 and 20 mm

At that point, it is defined the Safety Operation Area (SOA) where the implanted device can be positioned freely and still working properly because the recovered voltage is still sufficient to drive the circuits. Following the idea introduced before, the Safety Operation Area (SOA) reduces its size if the distance (Z axis) between both antennas increases.

The SOA can be observed in the XY plane view in Fig. 4.10. The recovered rectified voltage (V_{rec}) distribution is compared, for three Z axis distances, with the

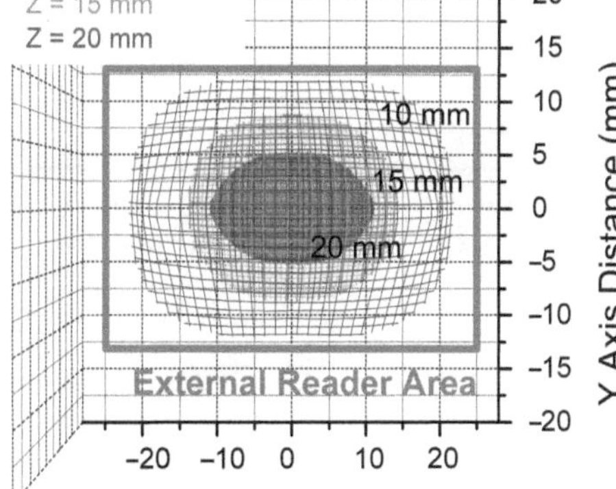

Fig. 4.10 Safety operation area (SOA) of the implantable device taking into account the XYZ misalignment using as external reader antenna the TRF7960

total available area of the external reader (orange rectangle). For a Z distance of 10 mm the SOA area is closed to the maximum area available by the reader. On the other hand, the SOA reduces till a minimum area of 20 mm × 10 mm for the highest height of Z = 20 mm (red zone in Fig. 4.10).

This situation is also illustrated in Fig. 4.7 where the External Reader Area (orange rectangle) of the TRF7960 is depicted and the obtained SOA for Z = 10 mm.

It is important to remark that the experimentally validated SOA is the area where the implantable event detector can be positioned and still working. Basically, the SOA area depends on the amount of transmitted power, size of the external reader, area and type of the implanted antenna and distance between them. It is important to remark these items.

The next step consists in validating the prototype working as a detector of electrochemical substances. The detection instrumentation and communication mechanism are evaluated using several concentrations of $K_4[Fe(CN)]_6$ in PBS and a commercial sensor AC1.W1.R1.

The implantable device is configured with a transimpedance resistance (R_{TIA}) of 25 k^{Ω} and a gain of 1 (R_G = 1 M^{Ω}). The programmed threshold voltages are 0.125, 0.250 and 0.500 V for V_{th1}, V_{th2} and V_{th3} respectively. A capacitor of 330 nF defines a communication time slot of 250 ms, the width of each transmitted zero. The final configuration of the prototype is summarized in Table 4.5.

First of all, the potentiostat circuit working together with the commercial sensor AC1.W1.R1 is verified. A cyclic voltammetry [41] from –1 to 1 V at 0.05 V/s is carried out using the external input of the device, pins S1 and S2 non-connected Table 4.2.

Figure 4.11 depicts the voltammograms obtained for 1 and 4 mM of $K_4[Fe(CN)]_6$. with the transimpedance resistance of 25 k^{Ω}. The results are compared with the ones obtained with the commercial instrument CH 1232A (CHInstruments®) and they are close to those presented in more detail in Section 3.3.3.

As can be observed in Fig. 4.11, the reduction and oxidation potentials are located around of 0.170 and 0.240 V. Moreover, the shape of the voltammogram is almost equal in both instruments with a maximum difference in the peak amplitude.

Table 4.5 Implantable detector prototype final configuration

Symbol	Parameter	Value	Units
C_{PRG}	Monostable capacitor	330 n	F
R_{TIA}	Transimpedance resistance	25 k	Ω
R_G	Gain resistance	1 M	Ω
V_{IN}	Programmable internal voltage Pins $S_1 = S_2 =$ Gnd	0.5	V
V_{th1}	Programmable threshold voltages Pins $S_3 = S_4 =$ Gnd	0.125	V
V_{th2}		0.250	V
V_{th3}		0.500	V

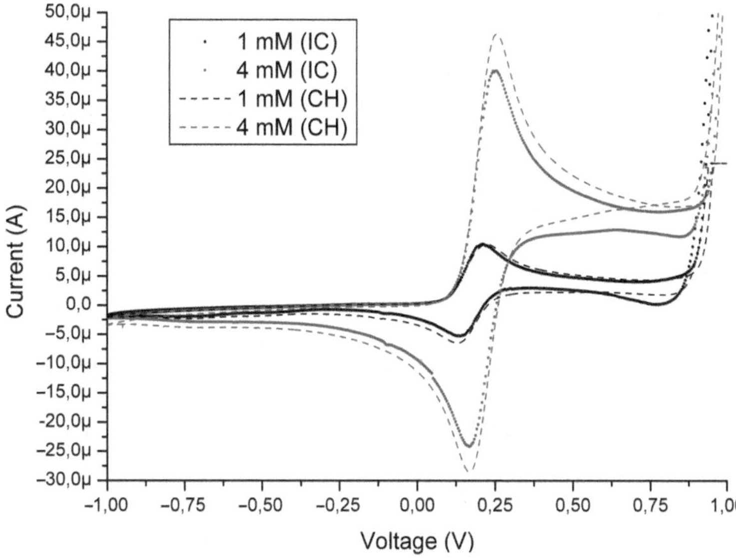

Fig. 4.11 Voltammogram obtained with the rectangular prototype device for 1 and 4 mM of $K_4[Fe(CN)]_6$ at 0.05 V/s

Once the correct operation is checked, it is configured to perform an amperometric analysis [41] to detect three concentrations of $K_4[Fe(CN)]_6$. In that way, pins S_1 and S_2 are grounded to internally generate 500 mV to be applied to the sensor.

Moreover, the Modulation and Data processing block is ready to detect three different currents around 6, 11 and 21 μA using $V_{th1} = 0.125$ V, $V_{th2} = 0.125$ V, V_{th3}. = 0.500 V. This relation is obtained by substituting the variables in eq. (3.7) for the values of Table 4.4 as follows:

$$V_{OUT} = I_W R_{TIA}\frac{R_G}{R_3} \Rightarrow \begin{cases} V_{th1} \leq V_{OUT} = I_W^{th1} 25\,k\Omega \rightarrow 0.125\,V \leq 6\,\mu A \cdot 25\,k\Omega \\ V_{th2} \leq V_{OUT} = I_W^{th1} 25\,k\Omega \rightarrow 0.250\,V \leq 11\,\mu A \cdot 25\,k\Omega \\ V_{th3} \leq V_{OUT} = I_W^{th1} 25\,k\Omega \rightarrow 0.500\,V) \leq 21\,\mu A \cdot 25\,k\Omega \end{cases}$$
$$R_G = R_3 = 1\,M\Omega \;/\; R_{TIA} = 25\,k\Omega$$

The detection points as well as the amperogram are presented in Fig. 4.12. The response is similar to those obtained with the commercial instrument CH 1232A and similar to those explained in detail in Section 3.3.3. As can be observed, during the transitions there are some current peaks that could produce errors in the detection; so the amperometric signal is filtered through an integrated 4th-order OTA filter to avoid false detections.

The communication mechanism is prepared to transmit 1, 2 or 3 zeros to the external reader in function of the achieved threshold point is V_{th1}, V_{th2} and V_{th3} respectively (more details in Section 3.5.1).

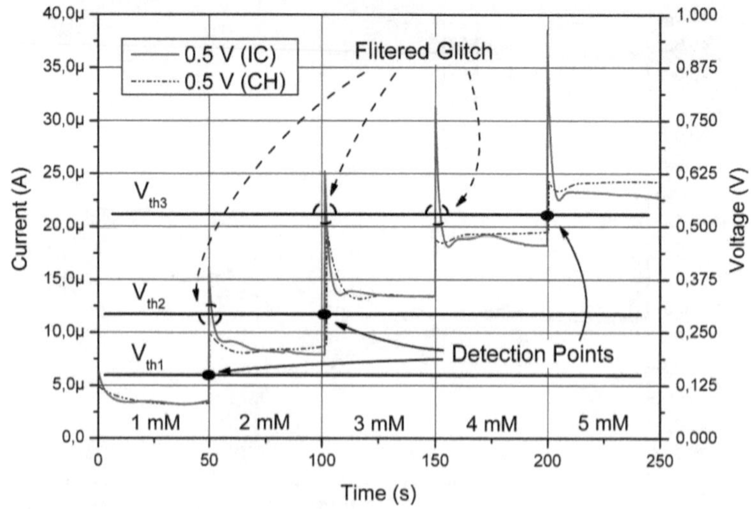

Fig. 4.12 Amperogram and programmed detection points using the threshold voltages th1, th2 and th3 and the equivalent current detected. The glitches produced in the transitions are internally filtered using the OTA filter

The actual prototype configuration defines a transmitted zero as a time slot of around 250 ms, ($C_{PRG} = 330$ nF, Table 4.5). Hence, a time slot of 500 ms has to be received in the external reader when two zeros are transmitted and 750 ms for three zeros.

Figure 4.13 depicts the waveforms of the internal and external antennas when all detection points are achieved. Plots (a) and (b) show the transmitted zeros for V_{th1} and V_{th2}. It is received at the external antenna a time slot of around 266 and 497 ms that corresponds to the transmission of one zero (V_{th1}) and two zeros (V_{th2}) respectively. These times are quite close to the expected ideal ones of 250 and 500 ms. In both cases, the communication is obtained for a distance of 10 mm between antennas at point (0.0). The modulation amplitude at the receiver side is 30%.

The communication of the third threshold point is depicted in Fig. 4.13c. In that case, the transmission is done for the worst case of 20 mm of separation between antennas. As can be observed, the AM modulation amplitude has decreased to 10% defining the limit of the communication distance range; the amplitude of the external signal is enough to detect the transmitted zeros. In that case, the time slot is around 780 ms, which corresponds to transmitting three zeros of approximately 250 ms each one.

Once the thresholds detection and communication are checked, the CMOS Front-End Architecture for In-Vivo Biomedical Implantable Detection Devices is fully validated. Finally, Table 4.6 summarizes the most important electrical features of the developed device.

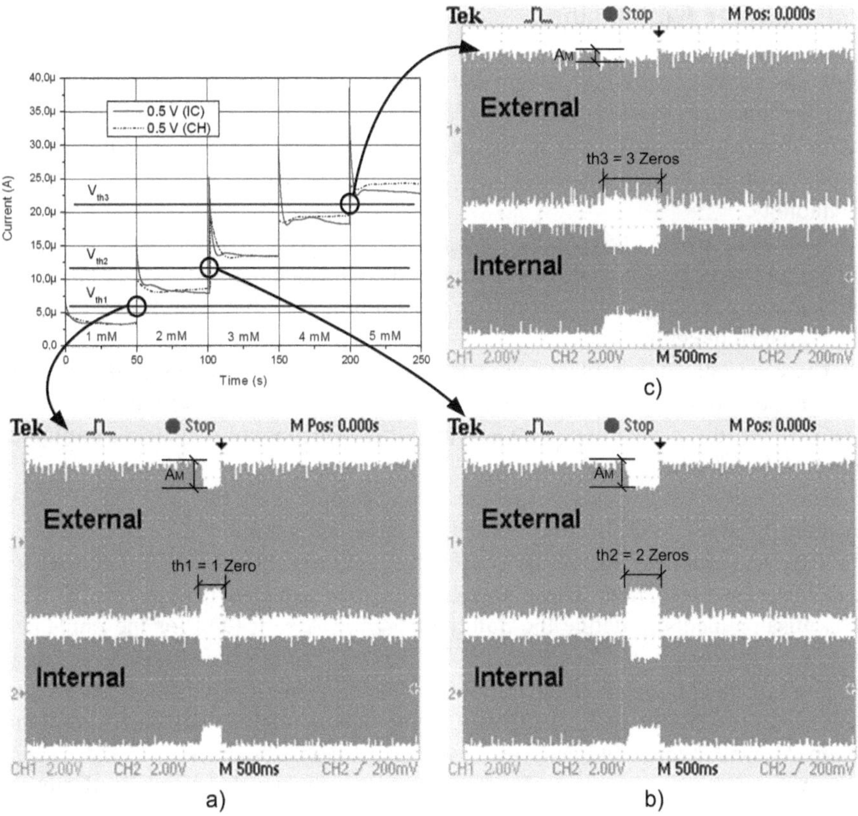

Fig. 4.13 **a** and **b** show the internal and external antenna waves when V_{th1} and V_{th2} are transmitted for a Z distance of 10 mm ($A_M = 30\%$) and (**c**) presents the antennas waveforms when V_{th3} is transmitted for the worst communication case of Z = 20 mm on air coupling ($A_M = 10\%$). The operation frequency is 13.56 MHz

Table 4.6 Implantable device main features

Symbol	Parameter	Min	Max	Units
V_{reg}	Regulated on-chip voltage	±1.12	±1.25	V
P_{rec}	Recovered power	800 μ	1.1 m	W
Z_D	Working distance between antennas	1	20	mm
I_W	Current detected	1 n	1 m	A
A_M	Data modulation amplitude at the receiver side	30	10	%
P_{CC}	Power consumption	700 μ	1 m	W
Freq.	Inductive powering and communication frequency	13.56		MHz

4.4 Chapter Conclusions

This chapter discusses the design and results of a front-end architecture detection system for in-vivo implanted devices. This solution combines in the same implantable package the communication circuit and antenna, power and sensor instrumentation modules, all of them fabricated in 130 nm CMOS technology.

Two small size prototypes have been fabricated as a first approach for an in-vivo implantable detection device. One uses a coiled antenna whereas the other one implements a rectangular antenna. The second prototype has been tested to check the whole architecture.

The electronics for implantable instrument is designed to be inserted under the human skin. It is oriented to work with three electrodes amperometric sensors thanks to the low-power biopotentiostat amplifier, which is able to detect a minimum current of +/–1 nA.

The electronics presents a low-power consumption of 680 μW @ ±1.2 V in order to avoid local heating dots. The maximum power that can be supplied through the inductive link, on air coupling, is 3.5 mW @ ±1.2 V (±1.5 mA) for a distance up to 15 mm. The power link has been studied in function of XYZ axis misalignments to define the Safety Operation Area (SOA) for the implanted device.

The instrument detection capability is validated using a commercial sensor and several concentrations of $K_4[Fe(CN)]_6$. Some voltammograms are carried out and compared with those obtained with a commercial instrument CH 1232A. Then, an amperometric configuration has been used to validate the threshold detection mechanism and the AM communication. In that way, the implantable detection architecture is experimentally fully validated.

Finally, a proof of concept was carried out with the coiled prototype to conclude with the prototype fabrication. It was covered with a lightly layer of biocompatible material (PDMS [42]) to explore a possible packaging solution for the in-vivo detection device. Figure 4.14 depicts this PDMS proof of concept prototype.

Fig. 4.14 Coiled prototype covered with PDMS

References

1. E. Ghafar-Zadeh, M. Sawan, Toward fully integrated CMOS based capacitive sensor for lab-on-chip applications, IEEE International Workshop on Medical Measurements and Applications, MeMeA 2008, (May 2008), pp. 77–80
2. D. Barretino, Design considerations and recent advances in CMOS-based microsystems for point-of-care clinical diagnostics, in *Proceedings of the IEEE International Symposium on Circuits and Systems*, (2006), pp. 4362–4365
3. Cygnus Inc, http://www.cygnusinc.net/
4. L. Cantarero, J. Butler, J. Osborne, The adsorptive characteristics of proteins for polystyrene and their significance in solid-phase immunoassays. Anal. Biochem. **105**, 375–382 (1980)
5. O.A. Sadik, A.O. Aluoch, A. Zhou, Status of biomolecular recognition using electrochemical techniques. Biosens. Bioelectron. **24**, 2749–2765 (2009)
6. M. Sawan, H. Yamu, J. Coulombe, Wireless smart implants dedicated to multichannel monitoring and microstimulation. IEEE Circuits. Sys. Mag. **5**, 21–39 (2005)
7. C.M. Zierhofer, E.S. Hachmair, Geometric approach for coupling enhancement of magnetically coupled coils. IEEE Tran. Biomed. Eng. **43**, 708–714 (1996)
8. C. Sauer, M. Stanacevic, G. Cauwenberhs, N. Thakor, Power harvesting and telemetry in CMOS for implanted devices. IEEE Trans. Circuits Sys. **52**(12), 2605–2613 (Dec 2005)
9. Y. Li, J. Liu, A 13.56 MHz RFID transponder front-end with merged load modulation and voltage doubler-clamping rectifier circuits, IEEE International Symposium on Circuits and Systems, (2005), pp. 5095–5098
10. K. Myny, S. Van Winckel, S. Steudel, P. Vicca, S. De Jonge, M.J. Beenhakkers, C.W Sele, N.A.J.M. van Aerle, G.H. Gelink, J. Genoe, P. Heremans, An inductively-coupled 64b organic RFID tag operating at 13,56 MHz with a data rate of 787b/s, IEEE International Solid-State Circuits Conference, 290–614 (2008)
11. A. Gore, S. Chakrabartty, S. Pal, E. Alocilja, A multi-channel femtoampere-sensitivity conductometric array for biosensing applications, 28th IEEE Engineering in Medicine and Biology Science Conference, 6489–6492 (2006)
12. M.R. Haider, S.K. Islam, M. Zhang, A low-power processing unit for in vivo monitoring and transmission of sensor signals. Sensors Trans. J. **84**(10), 1625–1632 (Oct 2007)
13. C. Sauer, M. Stanacevic, G. Cauwenberhs, and N. Thakor, Power harvesting and telemetry in CMOS for implanted devices. IEEE Trans. Circuits Sys. **52**(12), 2605–2613 (Dec 2005)
14. Y. Li, J. Liu, A 13.56 MHz RFID transponder front-end with merged load modulation and voltage doubler-clamping rectifier circuit, IEEE International Symposium on Circuits and Systems, 5095–5098 (2005)
15. K. Myny, S. Van Winckel, S. Steudel, P. Vicca, S. De Jonge, M.J. Beenhakkers, C.W Sele, N.A.J.M. van Aerle, G.H. Gelink, J. Genoe, P. Heremans, An inductively-coupled 64b organic RFID tag operating at 13,56 MHz with a data rate of 787b/s, IEEE International Solid-State Circuits Conference, 290–614 (2008)
16. M.R. Haider, S.K. Islam, S. Mostafa, Z. Mo, O. Taeho, Low-power low-voltage current readout circuit for inductively powered implant system. IEEE Trans. Biomed. Circuits Sys. **4**(4), 205–213 (2010). ISSN: 1932-4545
17. H.A. Wolpert, Use of continous glucose monitoring in the detection and prevention of hypoglycemia. J. Diabetes Sci. Technol. **1**(1), 146–150 (Jan 2007)
18. Medtronic Minimed Inc, http://www.medtronicdiabetes.com/
19. Abbot Inc, http://www.abbott.com/
20. J.D. Newman, A.P.F. Turner, Home blood glucose biosesors: A commercial perspective. Biosens. Bioelectron. **20**, 2435–2453 (2005)
21. M. Frost, M.E. Meyerhoff, In vivo chemical sensors: Tackling biocompatibility. Anal. Chem. **78**(21), 7370–7377 (2006)

22. M.W. Jung, D.W. Kim, R.A. Jeong, H.C. Kim, *Needle-type Multi-electrode Array Fabricated by MEMS Technology for the Hypodermic Continous Glucose Monitoring System*. in Proceedings of the International Coference of EMBS. (San Francisco, 2004), pp. 1987–1989
23. H. Nim Choi, J. Hoon Han, J. Ae Park, J. Mi Lee, Won-Yong Lee, Amperometric glucose biosensor based on glucose oxidase encapsulated in carbon nanotube-titania-nafion composite film on platinized glassy carbon electrode. Electroanalysis **19**(17), 1757–1763 (2007)
24. A. Erdem, H. Karadeniz, A. Caliskan, Single-walled carbon nanotubes modified graphite electrodes for electrochemical monitoring of nucleis acids ad biomolecular interactions. Electroanalysis **21**(3–5), 461–471 (2009)
25. J. Wang, In vivo glucose monitoring: Towards "Sense and Act" feedback-loop individualized medical systems. Talanta **75**, 636–641 (2008)
26. E. Lin Tan, B.D. Pereles, B. Horton, R. Shao, M. Zourob, K. Ghee Ong, Implantable biosensors for real-time strain and pressure monitoring. Sensors **8**, 6396–6406 (Oct 2008)
27. Positive ID/Verichip White Paper, Development of an Implantable Glucose Sensor, http://www.positiveidcorp.com/white-papers.html
28. S. Zimmermann, D. Fienbork, B. Stoeber, A.W. Flounders, D. Liepmann, *in Proceeding Internatioal Conference on Solid-state Sensors*. A microneedle-based glucose monitor: Fabrication on a wafer-level using in-device enzyme immobilization (Actuators and Microsystems, Boston, MA, 2003), pp. 99–102
29. A. Hassibi, T.H. Lee, A programmable 0.18-μm CMOS electrochemical sensor microarray for biomolecular detection. IEEE Sens. J. **6**(6), 1380–1388 (Dec 2006)
30. R.D. Beach, R.W. Conlan, M.C. Godwin, F. Moussy, Towards a miniature implantable in vivo telemetry monitoring system dinamically configurable as a potentiostat or galvanostat for two- and three-electrode biosensors. IEEE Tran. Instrum. Meas. **54**(1), 61–72 (Feb 2005)
31. M.R. Haider, S. Mostafa, S.K. Islam, A Low-Power Sensor Read-Out Circuit with FSK Telemetry for Inductively-Powered Implant System, in *IEEE Midwest Symposium on Circuits and Systems, MWSCAS*, (2008), pp. 450–453
32. J. Sacristán-Riquelme, F. Segura, M. Teresa Osés, Simple and efficient inductive telemetry system with data and power transmission. Microelectron. J. **39**(1), 103–111 (Jan 2008)
33. P. Vaillancourt, A. Djemouai, J.F. Harvey, M. Sawan, EM radiation behaviour upon biological tissues in a radio-frequency power transfer link for a cortical visual implant. 19th IEEE. Eng. Med. Biol. Sci. Conf. **6**, 2499–2502 (1997)
34. J. Colomer-Farrarons, J. Brufau, P. Miribel-Català, A. Saiz-Vela, M. Puig-Vidal, J. Samitier, *Power Conditioning Circuitry for a Self-Powered Mobile System Based on an Array of Micro PZT Generators in a 0.13 μM Technology*, IEEE Insternational Symposium on Industrial Electronics, (June 2007), pp. 2353–2357
35. A. Lasia. *Electrochemical Impedance Spectroscopy and Its Applications Modern Aspects of Electrochemistry*, vol. 32, (New york, Kluwer Academic/Plenum Publisher, 1999), Chapter 2, pp. 143–243
36. L. Yang, Y. Li, C.L. Griffis, M.G. Johnson, Interdigitated microelectrode (IME) impedance sensor for the detection of ciable Salmonella typhimurium. Biosens. Bioelectron. **19**(10), 1139–1147 (2004)
37. A. De Marcellis, G. Ferri, M. Patrizi, V. Stornelli, A. D'Amico, C. Di Natale, E. Martinelli, A. Alimelli, R. Paolesse, An integrated analog lock-in amplifier for low-voltage low-frequency sensor interface, International Workshop on Advances in Sensors and Interface, IWASI, (June 2007), pp. 1–5
38. D. Rairigh, A. Mason, C. Yang, Analysis of on-chip impedance spectroscopy methodologies for sensor arrays. Sens. Lett. **4**(4), 398–402 (2006)
39. A.E. Moe, S.R. Marx, I. Bhinderwala, D.M. Wilson, A miniaturuzed lock-in amplifier design suitable for impedance measurements in cells. Proc. IEEE Sensors **1**(24–27), 215–218 (AUTRICHE 2004)

40. Texas Instruments TRF7960 (Rev. E) on-line documentation, http://focus.ti.com/docs/prod/folders/print/trf7960.html
41. C.G. Zoski, Handbook of electrochemistery. Elseiber. (2007). ISBN: 0-444-51958-0
42. J.C. Lotters, W. Olthuis, P.H. Veltink, P. Bergveld, The mechanical properties of the rubber elastic polymer polydimethylsiloxane for sensor applications. J. Micromech. Microeng. **7**, 145–147 (1997). doi:10.1088/0960-1317/7/3/017

Chapter 5
Conclusions and Future Work

Abstract In this last Chapter the main electrical characteristics of the proposed CMOS architecture for the implantable, self-powered event detector device, are summarized. Future aspects in the development of a new prototype are envisaged by the authors.

Keywords Implantable event detector · Inductive powering

5.1 Conclusions

The developed approach for an implantable subcutaneous event detector device has been discussed in this work, analyzing in more detail the particular case where a rectangular antenna is used instead of a coiled antenna. The rectangular prototype has been fully validated obtaining the following characteristics:

- Small size, $5.5 \times 29.5 \times 0.5$ mm^3.
- Self-Powered through an inductive powering link.
- Low-Power consumption, 700 μW @ ±1.2 V.
- Low current detection capability up to 1 nA.
- Transcutaneous communications at 13.56 MHz.
- Focused on working with three amperometric Biosensors.

The devices are envisaged to be integrated subcutaneously, small size and low-power consumption, and to work with three amperometric Biosensor electrodes. The validation of the final device as a True/False or alarm level detector has been carried out through several electrochemical analysis like voltammetries and amperometries.

Moreover, several studies have been carried out in order to study the effects of the misalignments between the external and implanted antennas. A Safety Operation Area (SOA) has been defined as the area where the implantable device can be positioned freely and still working properly.

The architecture of each implantable device is divided in two main blocs:

1. the powering and
2. the instrumentation and communications.

J. Colomer-Farrarons, P.L. Miribel-Català, *A CMOS Self-Powered Front-End Architecture for Subcutaneous Event-Detector Devices*, DOI 10.1007/978-94-007-0686-6_5,

Regarding the instrumentation, a full-custom potentiostat is designed as a main circuit to work with the sensors. Then, the processing circuitry is conceived to detect three programmable threshold values and finally an easy low-power communication protocol has been implemented for transcutaneously transmissions thanks to an inductive link and a backscattering method.

The architecture also incorporates other subcircuits like a Bandgap reference circuit, LPF (Low-Pass Filters) and monostables. The LPF avoids problems regarding glitches and noise in the signal coming from the potentiostat and assures a correct detection. On the other hand, a monostable circuit and an external capacitor control the width of the transmitted bits. All these circuits have been integrated in the same IC called BioChip.

Regarding the powering issue, at this stage, the device is Self-Powered through an inductive link. Exploiting the idea of Self-Powered, a Multi-Harvesting Power Chip (MHPC) has been designed. It is prepared to work with several types of ambient energy sources to provide greater versatility to a generic application, Table 5.1. The use of alternatives power sources, harvesting, is the perfect choice for such application where the use of batteries is difficult, or a difficult maintenance, or a long life is required.

Table 5.1 MHPC maximum electrical circuits features

		Min	Max	Units
Vin Pzt	Piezoelectric	+/– 1	+/– 2.5	Vac
Iin Pzt		–	5	mA
Vin SC	Solar cell	1.3	2.5	Vdc
Iin SC		–	10	mA
Vin MI	Magnetic induction	+/– 1.3	+/– 2.5	Vac
Iin MI			10	mA
Vout SSD mode	Single storage device	1	1.2	V
Iout SSD mode		–	20	mA
Efficiency (η)		30	70	%
Vout MSD PZT	Multiple storage device	1	2.5	Vdc Programmable
SC		–	1.2	Vdc
MI		–	1.2	Vdc
Iout MSD PZT		–	10	mA
SC		–	10	mA
MI		–	10	mA
η MSD PZT		40[a]	80	%
SC		48[b]	85	%
MI		35[c]	75	%

[a]The efficiency is calculated using the piezoelectric Quick Pack QP40W from Midé Tech (Chapter 2).
[b]The efficiency is calculated using the 6 solar cells XOB17 from IXYS. (Chapter 2).
[c]The efficiency is calculated using a rectangular antenna of 30 mm × 15 mm, 220 nF @ 13.56 MHz (Chapter 2).

The MHPC is able to collect energy from:

- mechanical vibration,
- solar radiation and
- inductive link, all of them managed from a single IC.

The MHCP is designed to obtain high conversion efficiency, especially from mechanical vibrations. A novel low-power circuitry to harvest this type of energy is presented and validated. The solar energy is regulated by using a LDO whereas the magnetically induced energy with high-efficient AC/DC converters.

All circuits in both chips, BioChip and MHPC, are designed to assure low-power and, in some cases, very low-power consumption. The consumption is an important issue when developing circuits for implantable devices to reduce and to avoid the appearance of hot spots in the biological surroundings of the subcutaneous implants.

A discrete potentiostat was developed before initiating the integrated CMOS implementation to gain experience in the design and avoid problems in the integration. Moreover, this PCB based circuit has been also used to perform electrochemical experiments, cyclic voltametries and amperometries, and to gain also experiences in this type of analysis.

Talking again about the instrumentation for Biosensor, this book introduces theoretically the concept of Lock-In amplifier circuit architecture and its validation by simulations.

This circuit is used to detect the real and imaginary component of an input AC signal and it can be applied in electrochemical analysis to perform EIS analysis and to detect the impedance variation of a Biosensor. The circuit is based on two Synchronous Demodulated Channels to obtain two DC signal proportional to the real and imaginary components of the analyzed signal.

In summary, this book has presented a novel architecture for developing Self-Powered implantable detection devices and its validation through two prototypes. Table 5.2 presents the electrical characteristics of the MHPC. Table 5.2 summarizes the features of the validated implantable architecture. Thus, the incomplete pyramid presented at the introduction is now completed as is depicted on Fig. 5.1.

Table 5.2 Implantable device main features

Symbol	Parameter	Min	Max	Units
V_{reg}	Regulated on-chip voltage	±1.12	±1.25	V
P_{rec}	Recovered power	800 μ	1.1 m	W
Z_D	Working distance between antennas	1	20	mm
I_W	Current detected	1 n	1 m	A
A_M	Data modulation amplitude at the receiver side	30	10	%
P_{CC}	Power consumption	700 μ	1 m	W
Freq.	Inductive powering and communications frequency		13.56	MHz

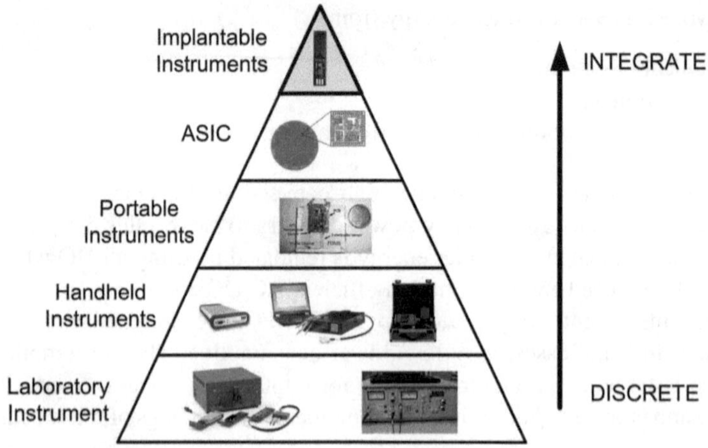

Fig. 5.1 Pyramid representing the evolution of biomedical diagnosis equipment with the achieved objective (top of the figure)

5.2 Future Work

Derived form the work presented in this book, different options would be opened. It would be possible to develop and integrate better instrumentation and electronic circuits in order to reduce the final size. It would be a nice evolution to integrate all the necessary circuits in one IC to mimic a grain of rice, Fig. 4.5.

The advances of the microtechnology and the channel length reduction in the transistors, there are available technologies of 90, 60 and 45 nm length that will help in the deployment of the subcutaneous low-sized implantable devices, but with high computational capabilities and mixed signal interfaces. So, controlling and minimizing the power consumption.

On the other hand, the development of architectures able to work with different Working electrodes (Multi-Sensors) at the same time would be an interesting evolution. In this way it would be possible to work with arrays of Biosensor and detect more than one substance, ion, ADN, etc ... at a time. Moreover, the development of powerful transcutaneous communication mechanisms to send such amount of information without increasing the power consumption of the subcutaneous device would be a must. At the end, a complex close-loop system, able to detect and actuate automatically, will be developed. For instance, the development of an artificial pancreas which would inject automatically insulin when some levels of glucose are exceeded.

Improving the powering of the device and developing better self-powered systems is also a very interesting research area. The integration of the MOS circuits and the MEMS microgenerators in the same substrate will allow the conception of more modern harvesting chips while reducing the final size of the application. Then, the possibility to use it in small, low-power and difficult maintenance system like in-vivo or in-vitro on-body systems is opened. Recently, research in the field of energy harvesting based on ultrasonic powering seams a new alternative to the inductive powering for implanted devices.

Appendix A

Bandgap Montecarlo Simulations

A statistical analysis has been performed with the reference Bandgap voltage circuit presented in Fig. 2.12. Two different supply voltages has been used, 1.4 and 1.2 V and 15 different integrated circuits has been tested.

The experimental results are compared with the expected theoretical values obtained by simulation with the statistical available models, fast (FFA), slow (SSA) and standard or typical (STD/TTY) of the technology.

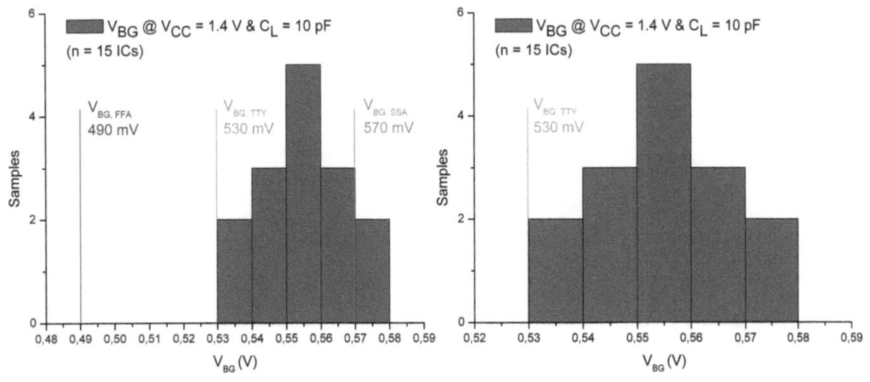

Fig. A.1 Experimental Bandgap voltage at 1.4 V. *Left* image shows the obtained values and the theoretical values and on the *right*, there is plotted in detail the experimental voltages

J. Colomer-Farrarons, P.L. Miribel-Català, *A CMOS Self-Powered Front-End Architecture for Subcutaneous Event-Detector Devices*, DOI 10.1007/978-94-007-0686-6,

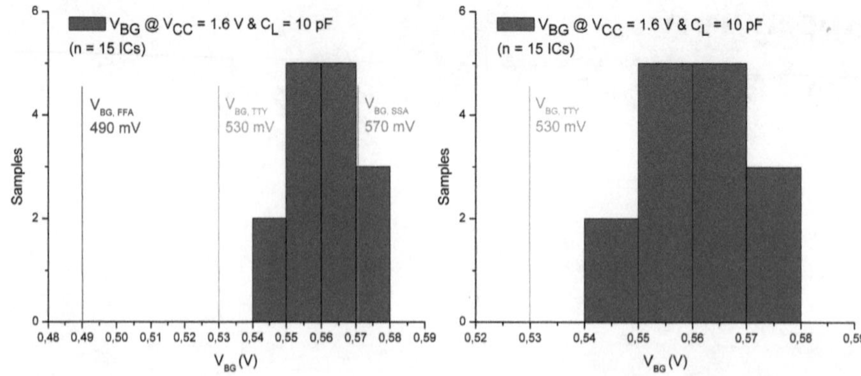

Fig. A.2 Experimental Bandgap voltage at 1.6 V. *Left* image shows the obtained values and the theoretical values and on the *right*, there is plotted in detail the experimental voltages

Appendix B

CMOS 130 nm Technology (HCMOS9)

The STMicroelectrics CMOS 130 nm technology provided by the CMP service (Multi-Project Circuits) is used to design and manufacture the IC's in this work. A brief description of the process specifications is done here. All the information can be found on-line at the official web page of CMP service, http://cmp.imag.fr/products/ic/?p=STHCMOS9.

- Gate length: 0.13 μm (drawn), 130 nm (effective).
- Triple well.
- Power supply of 1.2 V.
- 2.5 V power supply option is also available.
- Multiple Vt transistor offering (Ultra low leakage, High speed).
- Threshold voltages (for 3 families above):

  VTN = 570/500/380 mV,
  VTP = 590/480/390 mV.
- Isat (for 3 families above):

  TN @ 1.2 V: 410/535/680 uA/mic;
  TP @ 1.2 V: 170/240/320 uA/mic
- 6 metal layers in standard.
- Up to 8 metal layers in option.
- Low k inter-level dielectric.
- MIM capacitances.

Appendix C

CH Instruments CH 1232A Potentiostat

The CH 1232A is a commercial handheld potentiostat/biopotentiostat from CH Instruments. It is a part of the CHI 1200A series. Its main specifications are detailed below and can also be found on-line at the web page of CH Instruments, http://www.chinstruments.com/chi1200.html.

- Maximum potential range: $\pm$ 2.4 V.
- Compliance voltage: $\pm$ 7.5 V.
- Current range: $\pm$ 2 mA.
- Reference electrode input impedance: 1×10^{12} Ω.
- Sensitivity scale: 1×10^{-9} – 0.001 A/V in 7 ranges.
- Input bias current: < 100 pA.
- Current measurement resolution: < 5 pA.
- Data acquisition: 16 bit @ 10 kHz.
- CV and LSV scan rate: 0.000001–10 V/s.
- Low pass filter for current measurements.
- Maximum data length: 128–4096 k selectable.
- Power: UPS port of a PC.
- Chassis dimension: 7$''$ (W) $\times$ 4.5$''$ (D) $\times$ 1$''$ (H).
  19.05 cm (W) $\times$ 11.43 cm (D) $\times$ 2.54 cm (H).

CV = Cyclic Voltammetry
LSV = Linear Sweep Voltammetry

Kolibri 360 GmbH
Ferdinand-Jühlke-Straße 7
99095 Erfurt, Deutschland
produktsicherheit@kolibri360.de